BALANCE

BALANCE

A Dizzying Journey Through the Science
of Our Most Delicate Sense

Carol Svec

CHICAGO
REVIEW
PRESS

Published by Chicago Review Press Incorporated
814 North Franklin Street
Chicago, Illinois 60610
ISBN 978-1-61373-482-7

Library of Congress Cataloging-in-Publication Data
Is available from the Library of Congress

Typesetting: Nord Compo

Printed in the United States of America
5 4 3 2 1

For Bill, as always

Contents

Introduction

WHEN IT COMES TO BALANCE, ballerinas, circus performers, surfers, and gymnasts have nothing on you.

From the moment you wake up in the morning, daily life is a harrowing high-wire act. It's more than simply sitting upright or standing firmly on your feet without toppling over, although those are the most apparent aspects. To sit "upright" and not topple "over" requires having a solid sense of which way is *up*, so you can avoid falling *down*. It also means knowing where your body is in space and how to organize and move your muscles so that you can maintain your position. It requires a kind of internal gyroscope to keep the head level, a method for detecting gravity (so we don't fall victim to its pull), and a way to link what we see with how we move (so life doesn't look like a herky-jerky handheld video). Each step we take is a miracle of balance as our bodies compensate for hundreds of tiny changes in weight distribution and body position.

Without realizing it, we are—each of us (most of us)—a hair away from collapsing in a heap on the floor. I mean that literally. There are groups of tiny sensory hair cells in the core of the vestibular system within the inner ear that sense movement, acceleration, and gravity. When something goes wrong and those hair cells send the wrong message to the brain—as happens with Ménière's disease or benign paroxysmal positional vertigo (BPPV)—the result is incapacitating vertigo and dizziness.

But the sense of balance extends way beyond the inner ear. Nearly every cell in our bodies plays a role. Simply standing in place requires contributions from our eyes, ears, brain, nerves, joints, ligaments, blood, and bones in a symphony of coordination. Take away one factor and the body can usually compensate. Take away two and we may end up in a heap on the floor.

And yet, we don't think about this remarkable, exquisite ability we have until we find ourselves in unusual situations that stress the system, such as crossing an icy mall parking lot when you're overloaded with shopping bags. (I speak from experience.) All of a sudden, your firm footing is lost on the slick pavement and you have to think about each step, while simultaneously adjusting for the weight of your purchases. You plant a foot, pause, make a balance judgment, slowly lift your other foot until you know whether you're solid, take the step, plant the foot, and so on, until you reach your car. It's exhausting. And for a handful of people with a very specific proprioceptive disorder, a carpeted room is as challenging as that icy pavement.

So when you think about it, there's not much difference in physical skill required to walk a tightrope as there is in . . . well, just plain walking. I have spent the past year talking with physicians and researchers, visiting their offices and labs, learning about what "balance" really means. How do we accomplish this pure physical poetry every minute of every day?

What I discovered is that we don't *have* a sense of balance. We *are* balance. Balance gives us our place and space in the world, but it also contributes to our sense of self. After all, a dream of flying is impossible if you don't know which way is up.

I was inspired to write this book because of what happened to my mother. Back in December 2012, after a day of shopping, she lost her balance walking up the three little steps that led from the garage into her home. She fell backward onto the concrete garage floor, directly onto her head. Her only other injury was a slight scrape on the side of her arm, which flailed out in an automatic attempt to keep from falling. Just moments earlier, she had been lively, alert, and stable. But just three little steps and a quirk of balance later, she was unconscious.

I'm happy to report that my mom had a miraculous recovery (the neurosurgeon's words, not mine), despite skull fractures in two places and three areas of bleeding in her brain. We still don't know what made her lose her balance. Testing showed she did not have a stroke, had taken no medication or alcohol, and was—except for these most recent injuries—entirely healthy.

So why did she fall? And, perhaps more importantly, how come more of us *don't* fall? What keeps our balance steady? What happens when balance fails? Why are we so certain that we can come home from a trip to Bed Bath & Beyond and not end up in an ambulance? And what, exactly, does it take to knock us off our feet?

How to Use This Book

I want to make clear that I am not a medical doctor. (There probably isn't much risk of anyone making that mistake, but still.) This book was written to track my own exploration of the amazing world of balance research. It is not a compendium of disorders, and it should not be used to self-diagnose. If you recognize your own symptoms within these pages, please see a balance specialist, who can properly test, diagnose, and treat your very individual condition.*

The information in this book has been arranged in a way that builds knowledge, so that by the time you reach the last chapters, you can see how all the disparate and sometimes seemingly random bits of information come together. You can read the chapters in any order you like, but some of the concepts will make more sense if you read from the beginning.

After more than eighteen months of research, interviews, and writing, I have had two revelations. The first is that balance research is conducted by a tight-knit group of scientists with a high level of camaraderie and respect. The individuals you will meet in this book are among the top in their fields, yet they wanted to make clear that their work was part of a team effort. The second revelation is that my most overused phrase is "Oh, wow!" I have said it on the phone while talking with physicians and in person when I visited the labs of balance researchers. Most of the time, the scientists would reply with something along the lines of "I know! It's cool, isn't it?" It is a joy to write about scientific concepts when the experts are as excited as I am to talk about their work. My hope is that, at some point while you are reading, you also find yourself thinking, *Oh, wow!*

---

*For more information and support, visit the website for the Vestibular Disorders Association at www.vestibular.org.

1

Hurling for Science

Motion Sickness

S INCE THE 1960s, when there were no ethics boards to review
the designs of university studies using human subjects,
research psychologists have battled the suspicion that they are mad
scientists searching for the secrets of mind control. As professor
Andrea Bubka leads me down into her motion sickness laboratory,
it strikes me that she isn't doing her profession any favors. Well,
not Dr. Bubka herself—she has an open, cheerful face that I trust
at first sight. She's trim, petite, and with blonde hair ideal for an
academic—neither too long nor too short, neither too severe nor
too mussed, neither styled nor sloppy. Goldilocks hair: everything
just right. Plus she's wearing pearls. How can you not trust someone in pearls?

Rather, it is her lab that evokes fears of psychological sadism.
Apart from the bright lighting, it looks like something out of a
high-tech horror movie. It's a long bunker, with a gray concrete
floor, cinderblock walls, and industrial fluorescent lighting. Off

in the corner of the room is a wood and metal structure dangling a cylinder five feet tall and five feet in diameter. In the motion sickness research community, this device is infamous: the Vominator.

"Not everyone is happy with that nickname," Dr. Bubka explains. "Sounds too much like a carnival ride, not a serious piece of equipment."

I think it's kind of perfect.

The Vominator is officially known as an optokinetic drum. It is, indeed, a serious piece of equipment, although the one in front of me was homemade by Dr. Bubka and her research partner and significant other, Dr. Frederick Bonato. Their research line was inspired in the early 1990s after seeing what looked like a kind of rotating vinyl shower curtain at the Exploratorium in San Francisco, California. They first pieced together the Vominator in 2000 in an empty office, the creative sparks already flying. It turned out to be the key to a whole new area of research, and that's when they moved to the dungeon—uh, I mean basement.

They devoted the next fifteen years to bringing people to the brink of tossing their cookies. (That seems like a touch of "mad scientist," doesn't it?) The Vominator has made Drs. Bubka and Bonato famous in the field of motion sickness.

That's right—motion sickness, the most common balance disorder. You see, balance is about more than simply remaining upright; it's about predictability and steadiness in three main balance systems: the vestibular system (located in the inner ear); vision; and proprioception, or information from sensory nerves of the body. And few things shake up our equilibrium more reliably than a ride on a choppy sea, a spin on an amusement park ride, or the Vominator.

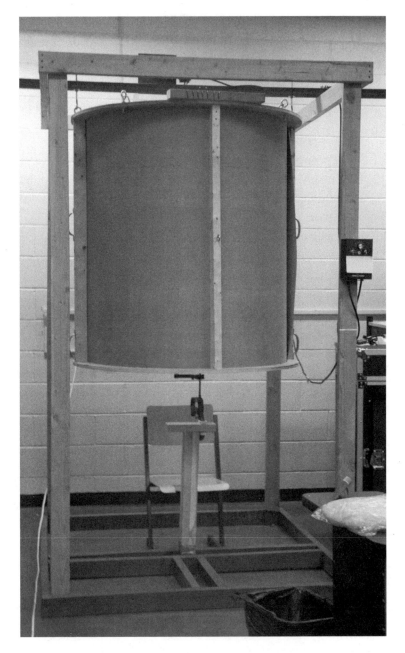

The "Vominator" optokinetic drum, viewed from the outside.

Carol Svec

I'm excited to see how it works, especially since I'm sure I'm going to be able to beat the device. I'm a tad competitive that way—I love technology, but I want to be able to control it. Machines should not (cannot!) control me.*

At the time I visited the lab, here's what I knew about motion sickness:

1. Some people are more susceptible than others.
2. I am highly susceptible, with a long and messy history of motion sickness. I get seasick standing on a dock, and if I go to an amusement park, I'm the one who waits on the side holding hats, handbags, and sunglasses while everyone else gets on a ride. A trip to Disney World is torture, mainly because I can't predict which of the ingenious rides will trigger my need to hurl.
3. Part of one's susceptibility to motion sickness is psychological. It is possible to make yourself sick just by believing it will happen. One bad experience on a roller coaster can trick your brain into thinking that all roller coasters will make you sick, and you're stuck on the sidelines forever. I offer myself as Exhibit One: When I was a young girl, my family had a speedboat that we enjoyed every weekend, all weekend. Then I turned fifteen and got a serious boyfriend. I wanted to spend my weekends with him, and I resented being dragged onto the boat for family fun. (I know, poor baby, right? What can I say, I was a teenager.) I developed very real seasickness that summer—I didn't just claim it; I experienced every nauseating minute of it. Eventually, I was allowed to stay home while my family sped away across the bay. It was a convenient excuse for me then, but I've been unable to step on a boat ever since. On the other hand, women are more sensitive to motion than men, a difference thought to be linked to hormones and physical development. So, whether my personal experience with seasickness was due to a desire to escape the boat or was the

---

*Except, of course, for my laptop, which has trained me to pay attention to it whenever I hear the ding of a new e-mail.

result of a more physiological process, I think we can all agree that my hormones were to blame.

4. If our psychology can create motion sickness, then maybe it can prevent it, too. Ergo, if I can convince myself I *won't* get sick, I'll overcome the queasiness and defy the Vominator.

That was my plan. I spent ten days visualizing myself feeling calm and centered. I looked at scientific research papers, YouTube videos, and even IMAX movies with ocean scenes (which usually make me feel slightly ill) to train my mind for the optokinetic drum. After all that preparation and positive thinking, it was with great confidence that I ventured into Dr. Bubka's lab to ask for a test run.

Inside the drum, I sit in a small chair. In front of me is a chin cup and forehead rest to keep my head steady, very much like the apparatus that holds your head during an eye exam. The entire inner surface of the cylinder from top to floor is a pattern of staggered black and white stripes:

Example of an internal grid from an optokinetic drum.
Carol Svec

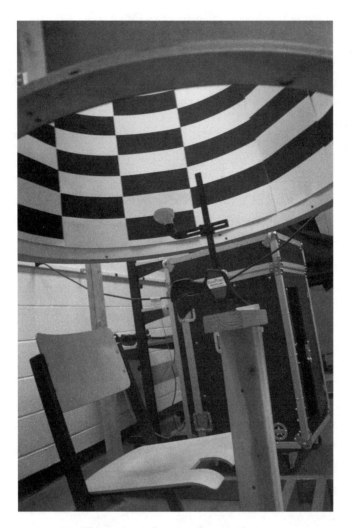

Closer view of an optokinetic drum.
Carol Svec

I am instructed to keep my head still and watch the stripes in front of me as the drum starts to turn slowly to the right. After just a few seconds, I see the drum stop moving and feel my chair—which is planted on the concrete floor—start to turn to the left. It's a vection illusion, a trick of the mind and the eye, but it's so powerful and so real that I grip the seat of my chair for balance and laugh out loud.*

Outside the cylinder, I hear Dr. Bubka say, "Ah, the chair just started moving, right? Let me know if you start to feel nauseated or uncomfortable."

Here's what I now know about motion sickness that I didn't know then: Almost everybody will eventually feel sick in the Vominator. Some within seconds, some within ten minutes. In my case, I went from feeling optimistic and assured to slightly green in less than three minutes—not a record but certainly on the sensitive side.†

The lesson is that when motion sickness strikes, there's no fighting it. Biology wins every time.

How We Get Motion Sick, or Why You Should Be Happy You're Not a Frog

If you have ever experienced motion sickness, the memory of those tortured hours are probably still quite vivid, no matter how long ago it happened. As the saying goes, "First you think you're going to die. Then you're afraid you won't." It starts with a general feeling of being unwell, perhaps a little agitation. Next comes any combination of dizziness, headache, confusion, drowsiness, and fatigue. Eventually your skin turns pale and may even take on a

---

*Have you ever been in a car stopped at a traffic light and had the car next to you start to move forward, but you perceive that *your* car is rolling backward? This is an example of a vection illusion.

†As with all her research subjects, Dr. Bubka released me from the optokinetic drum before I actually tossed my cookies, although I happily would have given a piece of myself to benefit research. No one has actually vomited while in the device—those pesky human research review boards forbid running the experiments to that ultimate conclusion. Plus, imagine how tough it would be to get graduate students willing to do the scut work in the lab year after year. However, a few people have reported being sick once they reached their car or even later at home.

greenish hue. Finally, the coup de grâce: cold sweats, accelerated heart rate, increased salivation, and incapacitation from nausea and vomiting. Beyond the physical discomfort, you actually can't think straight—concentration, reasoning, and reaction times suffer as much as your stomach.

What most people don't realize is that motion sickness doesn't necessarily stop when the motion does. Once you get off the boat or plane (or out of the Vominator), normal, garden-variety motion sickness symptoms can linger for up to three days.* You won't necessarily be nauseated, but you may continue to feel dizzy, foggy, tired, sleepy, or irritable. People who are particularly wrecked after air travel may actually suffer from a combination of jet lag and protracted motion sickness.

Some folks—like me—have a general, probably genetic susceptibility, but those who believe they don't get motion sick simply haven't been sufficiently tested. Research and real-time observations show that about 60 percent of military pilots get sick during their first flight, and up to 80 percent of astronauts are sick in the first three days in space. Sickness rates reach 100 percent for people in an ocean lifeboat and for those who go into an optokinetic drum. That's what makes it such an ideal scientific tool. Inside that checkered cylinder, it's nausea on demand.

Even those who are heavily trained can succumb if the conditions are right. Veteran sailors will get sick if the waves have a certain height and frequency. And on D-day, the paratrooper crew of the 101st Airborne Division was disabled with vomiting when the

---

*There is a less garden-variety form of motion sickness called *mal de débarquement*. This disorder gives a person the rolling, rocking, unbalanced feeling of being on a boat long after the initiating experience. How long? In some cases it can last years, or it can go away but spontaneously come back again. No one knows what causes it, but it is thought to be a problem in the way the brain tries to readapt to being on land after being on sea. Imagine developing sea legs that never go away.

planes had to dip and dive to dodge enemy attacks. In one plane, it started in the back with Cpl. Denver "Bull" Randleman, who vomited into his helmet. The next guy in line saw what happened and used his helmet as a barf bag, too. This began a chain reaction that continued throughout the plane. According to history books, the floor was awash in vomit, and some men forgot to empty their helmets before preparing to make the jump.*

It is also possible to have a kind of sneaky motion sickness known as sopite syndrome. About 60 percent of us do. With sopite syndrome, you don't even know you're sick because there's no nausea or vomiting. Instead, on a long drive or trip, you find yourself becoming listless, fatigued, distracted, and very drowsy. Scientists believe that when an otherwise healthy and well-rested person falls asleep while driving, it's probably due to sopite syndrome. After a very long while, you may even feel depressed and apathetic—typically enthusiastic and outgoing people become sullen, uncooperative, and withdrawn. For most of us, this is a minor inconvenience. But sopite syndrome is a significant problem for all branches of the military, which need troops to be alert and dependable at all times. As with regular motion sickness, these symptoms can last two or three days after finishing a journey. Only very astute medical professionals—and now you—can correctly diagnose this condition because it can seem like the sufferer is simply "in a mood" or coming down with a mild bug. (Don't worry; give it three days and your loved one will be back to normal.)

The only people who seem totally immune to motion sickness are those without a functioning vestibular system—which can be

---

*This demonstrates another interesting facet of motion sickness: Hurling is contagious. Scientists say we vomit when others do because we have "mirror neurons" in our brains that make us react with empathy when we see extreme emotion or behavior, and we mimic what we see. Plus strong smells can trigger nausea in some people.

a result of bilateral damage to the inner ears—and babies. Until age two, children do not experience motion sickness. The common reasoning is that until they learn to be steady on their feet, they need to be able to tolerate unexpected motion as they are carried, rocked, flipped over for changings, or dangled in a swing. If children felt symptoms whenever they stood, crawled, rolled, or fell, motion sickness would halt their development.

Humans aren't unique in their susceptibility—nearly all animals with a backbone (i.e., vertebrates) experience motion sickness, although they express their discomfort differently. Instead of vomiting (although many do that, too), animals may drool, pace, yawn, whine, become either agitated or lethargic, or develop diarrhea. Dogs are very close to humans in their response to motion, both in terms of sensitivity and reactions. Squirrel monkeys experience motion sickness so similarly to people that they are now the preferred test animal for scientists looking for a cure. Cats express their sickness by yowling. Horses do well traveling in a trailer but can become sick on a ship (although they can't throw up). Motion sickness has also been documented in seals, gerbils, pigs, sheep, cows, and birds. Even fish seem to experience motion sickness; when transported by plane, they appear disoriented and swim in circles—surrounded by bits of fish vomit.

The most heroic of motion-sick animals has to be the frog, more specifically, the adorable, bright green Japanese tree frog, which was the focus of the Frogs in Space program.* Seriously, that's what it was officially called. In 1990 six frogs were taken to the *Mir* space station for eight days. When allowed to float in the cabin, the frogs took up a "parachuting" posture, with all four limbs

---

*If you read that and immediately thought of the Muppets show *Pigs in Space* and heard the announcer's voice in your head, you're not alone.

outstretched, as if they were flying squirrels leaping from tree to tree. When perched, they sometimes appeared to retch, and they actually walked backward—something not typically seen in nature. In other words, their brains got pretty scrambled.*

The Queasy Side of Balance

To the nonscientist, motion sickness seems straightforward: bumpy rides make for queasy stomachs. Simple, right? The reality is much more complex and multidimensional. The truth is that even after centuries of study, researchers can't say for certain why motion sickness happens or what its physiologic mechanisms are. Why does motion—or even the illusion of motion, which is what's going on in the Vominator—cause us to become dizzy, confused, and nauseated? The downstream expert explanations have huge, rather important gaps. It's like watching an episode of *Seinfeld*: the body senses motion, relays signals to the brain, then yadda-yadda-yadda the next thing you know you're feeling sick. But what happens during the yadda-yadda-yadda? That's the question that keeps motion sickness researchers up at night.

For answers, I turned to Dr. Robert S. Kennedy, one of the pioneers of modern motion sickness research, a master of his field. He spent twenty-two years in active duty in the navy as a human

---

*You may wonder how scientists test animals for motion sickness. Sure, some of the observations are serendipitous, such as noticing fish behavior during a flight. But scientists who want to understand physiologic mechanisms need a reliable way to nauseate animals. After much trial and error, they discovered one secret: miniature Ferris wheels. OK, so they aren't brightly painted, nor do they come with calliope music, but they revolve slowly around a horizontal axis, with a counterbalance added to keep the "cars" upright. Apparently, these contraptions make animals really sick, especially cats and sometimes smaller mammals, such as mice and shrews. Other devices include a shaker plate, which simulates an earthquake, and a horizontal sled, which moves an animal sideways. Tell me that doesn't sound like a Tiny Town amusement park (cotton candy not included).

factors psychologist, doing research on how to prevent and treat motion sickness in a military environment. Thanks, in part, to Dr. Kennedy, we know about the unique problems of balance in divers, the best ways to land safely on a carrier at night, and how extended use of electronic visual displays—computers, iPads, video games—can distort our equilibrium (more about cybersickness in chapter 9). Dr. Kennedy was one of the inventors of the famous rotating room and is now an international human factors consultant. He has worked with such heavyweight clients as NASA, the US Navy, Martin Marietta, and Disney (those rides are a rich field for the study of motion sickness). He literally wrote the manual on how military personnel can avoid motion sickness when training in a simulator.

I had been warned that Dr. Kennedy never joked and wasn't too friendly to people he didn't know. "Just don't ask stupid questions," Dr. Bubka said, "and you'll be OK." Well, that's absolutely the quickest way to make a journalist feel self-conscious. I mean, stupid questions are kind of my thing—they often elicit surprising answers, so I like to sprinkle a few into any interview. I sweated over my list of questions for two days before I had a chance to speak with Dr. Kennedy on the phone, but I relaxed when I pulled up his photo online. He looks like Santa Claus in a well-tailored business suit.

It was true—Dr. Kennedy never once joked or chuckled at anything I said during our hour-long conversation, and believe me, I tried. But I understand it; he's a busy guy. Dr. Kennedy has published more than six hundred scientific articles in the past thirty years. That's one or two articles every month for the life of his career to date. It's not humorlessness but focus. I learned more in that one call than I had in weeks of researching on my own. I began with stupid question number one: Why do we get motion sick?

"We don't really know." Dr. Kennedy said. "When we study a phenomenon like motion sickness, we want it to be a *single* thing, *caused* by a single thing. But motion sickness has multiple expressions and multiple causes. Theories are simply a means to advancing understanding. All current models offer something valuable." His deep, resonant voice gave everything he said a weight of importance. He could make a good living doing "In a world . . ." narrations for movie previews. I want him to read me Edgar Allan Poe stories out loud.

He explained that of the top three theories of motion sickness, the overstimulation theory is the oldest. Before there were cars, trains, subways, airplanes, and spacecraft, there were boats, and where there are boats, there is seasickness. Now there is also amusement park ride sickness. Oh, and motion sickness is also common among novice riders of elephants or camels. (Dr. Kennedy—who has logged more than three thousand hours in flight simulators, including the zero-gravity "Vomit Comet" simulator—tells me that the sickest he has ever been was while riding a camel. "If you ever have the opportunity," he says, "skip it.")

The common denominators for all those varied modes of transportation are undulation and sway. If you are sitting in a chair, you are still except for motions you choose to make. On a boat, camel, or other sick-making way to travel, your body is subjected to lurching, tilting, bouncing, shaking, and wobbling. The overstimulation theory suggests that all the unpredictable, uncontrollable movements make the nerves in the inner ear, stomach, and other body organs fire like crazy. This overload is a signal to reduce attention to unnecessary functions (such as thinking clearly, apparently), while essentially allowing the stomach to freak out.

While that theory intuitively defines most of our own experiences, it doesn't explain visually induced motion sickness, like why

I became nauseated sitting perfectly still inside the rotating drum of the Vominator.

The next major theory to come along was the toxin theory, which, boiled down, is this: When animals and people eat something poisonous, symptoms often include dizziness, vertigo, or hallucinations, and the body's natural response is to expel the poison by vomiting. We become motion sick because the brain, trying to figure out why we are feeling or sensing motion (even though we are sitting still), decides that we're hallucinating, perhaps because we have eaten something poisonous. Again, shut down the thinking apparatus, increase power to the stomach. Plus, if you're feeling sick, you're going to hunker down, curl up in the fetal position, and rest, which slows down blood circulation, metabolism, and other body function that could influence how quickly any suspected poison might overcome the body.

While the toxin theory is attractive, it supposes that the brain can be easily fooled. If I don't eat anything before getting on a boat, I will still get sick. Why would my brain suddenly forget that no food was ingested? Shouldn't it have better communication with my digestive system? If I know I'm sitting still and the rotating drum is moving around me, why doesn't my brain know? Rather a large limitation.

On to idea number three, the widely accepted sensory conflict theory. If what you *see* and what your body *feels* don't match, there is conflict between two or more sensory systems, resulting in motion sickness. For example, if you're reading while riding in a car, your eyes see a stable, fixed environment (both the book and the interior of the car), but your body senses the car's motion. Those differing and conflicting signals cause confusion in the brain. The eyes see one thing, the body feels another, and the brain cannot reconcile the discrepancy. It does not compute. The result is sickness.* And the greater the magnitude of conflict—that is, the larger the difference between what is perceived and what is real, or between what two different senses detect—the sicker we get.†

The sensory conflict theory works to explain what happens with visually caused motion sickness, such as what happens in the Vominator. The body is not moving, and there is no expectation of motion, but the eyes show movement—ergo conflict, leading many people to become just as sick as if they were on a speedboat. But again, there's a piece missing: we have sensory conflict, yadda-yadda-yadda, now we're heaving over the rail.

For this particular example, science has filled in most of the gap. Our sensory systems, the eyes in particular, are keenly sensitive to motion. This response is hardwired. On a primitive level, it's what allowed our ancient ancestors to hunt, to track prey so they could eat, and to see and avoid any creature that might be moving in to eat them. When we see movement, it's exciting to us, and we can't look away. So it's natural for us to track movements, our eyes scanning from left to right, then jumping back again as the object moves out of our field of vision. Smooth tracking right, followed by a snap back to the left—smooth right, snap left—right, left, right, left. The connection between eye movements and balance is direct and powerful.‡

So, the Vominator causes motion sickness by exploiting the visual and vestibular system connection, tricking the brain into thinking that we are on the move even though we are sitting still,

---

*For all my fellow *Star Trek* fans (the original TV series, of course): This theory reminds me of the computerized probe called Nomad, the one that went on the fritz and destroyed itself simply because Captain Kirk fed it information that conflicted with its programming.
†Of course, extreme levels of conflict could cause your brain to just let go. Instead of trying to resolve the conflict, it would just ignore one stimulus or the other.
‡This eye-ear connection is explored much more deeply in the coming chapters.

walled off from the world by a spinning room. Sensory conflict creates motion sickness.*

More than Just Sadism at Work

You might be asking yourself, as I did, what purpose optokinetic drum research really serves. Once scientists discovered that sitting in a rotating cylinder makes people nauseous, what more was there to learn? As it turns out, quite a lot.

"Right now, we're testing the boundaries of what makes motion sickness worse," says Dr. Bubka. "Inside the drum, we started out with the standard black and white stripes. Then we moved to black-and-white checkerboard." That was the version I saw.

"When we add brightly colored stripes to the wall pattern, people get sicker faster," says Dr. Bubka. "The more colors, the sicker they get. The more complex the pattern, the sicker they get." So the greater the sensory bombardment, the bigger the conflict and the sicker we feel. At one point, researchers even added a wobble to the spin, which dialed motion sickness levels up past eleven.

A spinning drum in a lab may seem quite remote from daily life, but its application is evident all around us. Your car's interior design was influenced, in part, by sensory conflict theory. Whenever you have what is basically a closed-in box in a moving environment—like a car or a cruise ship cabin—motion sickness is a risk. Because of what scientists learned from optokinetic drum research, designers know to use blander color schemes and less complex patterns to help minimize the risk of sickness. In these cases, dull or

---

*Way, way back in time, long before engines, airplanes, and rotating research drums, people would experience this form of motion sickness via less technical moving visual stimuli, such as watching the rotation of a potter's wheel or the constant flow of a river.

monotone is good, vibrant is nauseating. Lesson learned: something positive can come out of even the bleakest laboratory.

But there are also uses for the flip side of the research. For all the Dr. Bubkas out there looking to reduce your nausea and discomfort, there's someone who wants to use that research to make your head spin. Literally. Welcome to the topsy-turvy, spinny-pukey world of amusement park rides, where engineers throw all the motion sickness no-nos at you at once. Bright colors. Complex patterns. Flipping, tilting, rotating, undulating carriages.

The parks themselves are designed for maximum adrenaline and sensory engagement with the environment. Lots of movement to catch our eyes. Everything is lit up with color. The patterns are all complex. And let's not forget the movement! Some people get sick just by *watching* a carousel (which, when you think about it, is similar to the experience of an optokinetic drum, so it's not surprising).

We love rides because they are exhilarating . . . in short doses. Most are designed to last less than three minutes specifically to avoid sensory conflict overload. If you want to get the most enjoyment out of amusement park rides, turn back the clock and become a teenager again. The peak time for motion sickness is between about ages six and twelve. For reasons that are not well understood, motion sickness fades after puberty, so by about age fifteen or sixteen, you can thrill to the excitement and overstimulation without nausea. But in adulthood, the thrill turns sour again. Why? No one knows.* If you used to live to ride roller coasters but now can't stomach them, consider it just another sad fact of getting older.

---

*A few more demographic details that scientists don't understand and can't explain: Women are more prone to motion sickness than men, and women are most sensitive during the week they are menstruating. Most optokinetic drum researchers will turn away menstruating women because their ultrasensitivity will skew the data. (Something

While rides are not designed specifically to increase the vomit factor (no matter how much it may feel that way), engineers don't attempt to eliminate motion sickness, either. I asked several ride engineers what variables they manipulated to control nausea and other motion sickness discomfort. They all laughed. Every single one of them. Motion sickness is nowhere on their working radar. One engineer who used to work for Disney told me that when a ride caused an excess number of guests to vomit (as opposed to, I suppose, the *usual* number of hurlers), the park's answer was to put up warning signs and provide barf bags.* Rather than serving as a deterrent, that "fix" just attracted more eager riders.

For the most nauseating time at an amusement park, look for the rides that break the most motion sickness rules. A good example is the Mad Tea Party, known to most Disney parkgoers as the spinning teacups. This ride has three small turntables that rotate clockwise, all set on a larger turntable that rotates counterclockwise. Each of the small turntables has six teacup cars that can be made to spin individually by turning a center wheel. Got that? You have spin, counterspin, and triple spin. Then add colors (bright) and designs (complex). Each cup is painted with a different

---

to keep in mind if you're planning a cruise or trip to an amusement park.) Women are also more likely to suffer migraines than men, and people with a history of migraines are much more motion sensitive than nonmigraineurs.

*I was delighted to discover that about one hundred people in the world collect vomit bags, or airsickness bags, as the more genteel call them. The world record holder is Niek Vermeulen of the Netherlands, with more than seventy-five hundred unique bags. The biggest collector in the United States is Bruce Kelly in Alaska, with more than six thousand bags. The best online collection belongs to Steve "Upheave" Silberberg (www.airsicknessbags.com), who believes that they are a form of art. Spend a few minutes on his site and you'll discover he's right. Sure, there are endless white bags with airline logos, but there are also gems, like the Finnaviation bag, a blue-and-white graphic masterpiece of a reindeer blowing chunks, or the Air Midwest bag, which features a zip closure, giving you the option of filling the bag with air, closing it, and using it as a makeshift pillow. When I asked Steve what made him start his collection, he told me that "chicks dig barf bags." Really? "No, not really. But maybe, if I say it often enough, it will come true."

pattern, and the floor disks have psychedelic swirls. It's no wonder this ride has a reputation for being among the worst for parents and grandparents to suffer through.

Despite everything we know about motion sickness, there is still no way to reliably prevent or cure it. The medications we use most often, namely dimenhydrinate (such as Dramamine and Driminate) and scopolamine, are very helpful for many people, but they are not universally effective. And while those medications might calm a queasy stomach, they don't do much to help all the other symptoms that can make you feel miserable. Plus those medications have a lot of side effects, including drowsiness and fuzzy-headedness, which aren't bad if you can just fall asleep in the backseat of a car for a couple of hours but can be debilitating during a ten-day ocean cruise. Motion sick pilots in the military, who can't afford to nap in a cockpit, are routinely given Dramamine and the stimulant dextroamphetamine to offset the sleepiness. (Keep in mind that they are professionals who need to stay alert during training and battle; please don't try this drug cocktail on your own.)

What does seem to work, according to experts, is keeping busy.

"The best treatment is distraction," says Dr. Bubka. "Astronauts and pilots who are sick are encouraged to do their jobs. We don't know why this works, but it does. Find a way to occupy yourself. Do a task. If that's not possible, look away or close your eyes. That removes one of the conflicting signals to your brain. If you close your eyes in the optokinetic drum, the nausea goes away."

2

Loops and Rocks in Your Head

The Vestibular System

ALL ORGANISMS NEED A WAY to control and maintain their alignment within their environment, both while moving and when standing still. Try to think of an example of a critter that might not need a sense of balance. What popped to my mind was the mushy-bodied, blobular, no-brained jellyfish. I imagined that they simply float along in the ocean, pushed around by tides and currents. Passive, helpless creatures. Boy, was I wrong! Apologies both to jellyfish and to the scientists who study and love them.

I spoke with Dr. Angel Yanagihara, director of the Pacific Cnidaria* Research Lab and research professor at the Békésy Labora-

---

*Cnidaria is the group of ocean creatures that includes jellyfish, corals, anemone, and hydroids—anything with a poisonous stinger. It was a painful and dangerous encounter with a box jellyfish that led Dr. Yanagihara to her expertise. While swimming off a Hawaiian beach, she was stung and barely made it back to shore before passing out on the sand. The experience changed her life, and she is now one of the world's leading experts in box jellyfish behavior and venom.

tory of Neurobiology and Department of Tropical Medicine at the University of Hawaii in Honolulu. Here's what I learned: Jellyfish are not just amorphous bags of stinging pain. They have definable structures, some of which allow them to sense gravity so they can tell which way is up or down, and how their bodies are tilted in reference to the surface of the ocean.

All species of box jellyfish have four eyestalks, one on each of the four sides of the box-shaped jelly. Each eyestalk has six eyes, one pointing up and others pointing forward. Those twenty-four eyes give the jellyfish essentially a 360-degree panoramic view. Suspended from the lower tip of each stalk, under the eyes, is a crystal about the size of a grain of sand, surrounded by additional crystalline layers.* These heavy crystal structures weigh down the eyestalk, giving jellyfish a method for detecting gravity so they can orient themselves within their environment.

Oh, and that crack I made about jellyfish being brainless? I was just demonstrating my own vertebrocentric ignorance. As Dr. Yanagihara points out, just because jellyfish don't have a central nervous system connected to a single computing center doesn't mean they don't have brain power.

"Perhaps jellyfish are redefining what it means to have intelligence," says Dr. Yanagihara. "Maybe each eyestalk has a neural computing center. It's true, there is no central brain, but jellyfish can make behavioral changes based on visual or gravitational inputs."

So, yes, even jellyfish have balance organs.

Our own balance needs are intricate and much more complicated. No matter which way our head is turned, regardless of the direction and speed of our movements, we have to be able to

---

*A new layer of crystal is grown for each day of a jellyfish's life, so, as with the rings of a tree trunk, scientists can determine the age of a jellyfish by counting the layers around the central grain.

distinguish up and down, forward and backward, left and right, even while running zigzags, jumping up and down, spinning in circles, or turning cartwheels. That ability starts with our vestibular system.

Of Peas, Gelatin, and Beer

Think of a pea—an average-sized green pea, one-third of an inch in diameter. That tiny sphere is large enough to contain the structures that form the core of our sense of balance: the vestibular apparatus.

There are two main parts of the vestibular apparatus: the semicircular canals and the otolith organs. The semicircular canals are three tiny, looping tubes set at nearly perfect ninety-degree angles from each other in three-dimensional space. They are the body's main motion sensors, detecting angular acceleration (also called rotational acceleration), which is a fancy way of saying that they sense all the ways a body can move other than up, down, forward, or backward. When you do a cartwheel, turn around to talk to a person in line behind you, or bend to examine your knees, your semicircular canals register the movement and allow you to keep your balance.

The three semicircular canals are set at different angles so they can detect different types of rotational movement. The superior canal and posterior canal are each oriented about forty-five degrees off a vertical midline plane. Place a finger just behind your right ear and trace a line to the outside edge of your left eye. That is the orientation of the right posterior and left superior canals.* If you place a finger just behind your left ear and trace a line to the outside edge of your right eye, that is the orientation of the left posterior

---

*The semicircular canals on the left side of the head are a mirror image to those on the right. Given the canals' orientations, this flips the relative positions of the posterior and superior canals.

and right superior canals. Finally, place a finger on one ear and trace a path across your face and over your nose to the other ear. That's the orientation of the smallest loop, the horizontal canal.

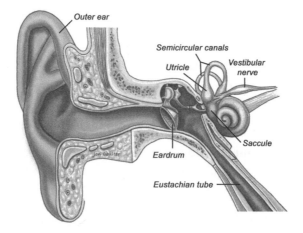

Overview of the anatomy of the ear.
Jon Coulter, M.A., CMI

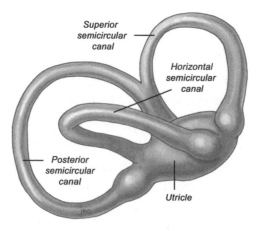

Close-up illustration of the semicircular canals.
Jon Coulter, M.A., CMI

Together these loops allow us to sense forward and backward rotation of the head, like when you nod yes or do a somersault; rotation of the head from side to side, like when you tilt your head trying to touch your right ear to your right shoulder or turn a cartwheel; rotation of the head left and right, like when you shake your head no or turn to see who tapped you on the shoulder from behind; or any other angle of motion.*

The semicircular canals contain endolymph, a fluid slightly thicker than water. It's about the consistency of an average beer. Move your head in any direction and the endolymph in the associated semicircular canal flows in the opposite direction. Each loop of the semicircular canals has a bulged end called the ampulla, which contains a structure called the cupula. The cupula looks rather like a flower bud and is filled with an almost jellylike substance and a bunch of miniscule, sensitive hair cells. These hair cells are key to keeping balance because they send positional information to the large vestibular nerve, which in turn feeds that information to the brain.

Here's what happens: When you move your head in a particular direction, the beer-like endolymph inside the associated semicircular canal pushes on the cupula, which bends and sways like a cone of gelatin wiggling on a plate. As the cupula moves, the tiny hair cells inside also get pushed around. This movement is detected by the vestibular nerve, which sends the information to the brain's balance center. In the brain, the signals are decoded and the position of the head is registered. This all happens virtually instantaneously.

---

*In the early 1800s, when scientists were trying to discover exactly what the semicircular canals did, they deliberately destroyed one or more canals in animals, especially, for some reason, pigeons and rabbits. If the damage was to the superior semicircular canal, the animal turned continuous somersaults. Damage to the horizontal semicircular canal caused the animal to spin in circles. Those early vestibular scientists had no idea that they were causing permanent vertigo in the animals.

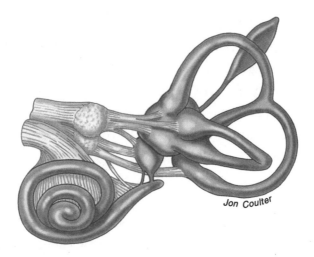

Close-up of how the large vestibular nerve connects to each individual
part of the vestibular system of the inner ear.
Artist: Jon Coulter; Reprinted from Earsite.com, with permission

Imagine tripping on a curb. Before you can consciously think, *Hey,
I just tripped on a curb,* your semicircular canals have already sent
the message of imbalance to the brain, and the brain has responded
with signals to your body about how to regain balance and keep
from falling on your face.

Semicircular canals can fire off signals individually or in any
combination. The vestibular nerve is capable of receiving and
combining multiple bits of information. With a perfectly functioning vestibular system, you can continually move your head
in any position and always know which way is up and be able to
keep your balance. No matter how quickly you move your head,
no matter how many times you turn and tuck, your brain keeps
track of which way is up. That processing speed is important.
Too much lag time and we would literally lose our place in the
world. Think of an elite gymnast doing a floor routine—despite
multiple flips, twists, and aerial gyrations, the athlete will always

know how to land on her feet. That's the role of the semicircular canals.

Our Bedazzled Inner Ears

With their graceful loops, the semicircular canals are the fancy, showy half of the vestibular system. They connect to a bean-shaped body called the vestibule, home to the other half of the vestibular system: the utricle and saccule (also known as the otolith organs). These are patches of tiny hair cells located on the walls of the vestibule, similar to the hair cells in the cupula of the semicircular canals. Each patch of tiny hair cells is covered by a gelatinous layer and blanketed by a membrane embedded with heavy crystals called otoconia.* In microscopic photographs, these "ear rocks" are beautiful—elongated crystals mounded in a jumble on top of the hair cells of the utricle and saccule. Each of these balance crystals is tiny, between just 3 and 30 microns long, or about one-tenth the size of a single crystal of table salt.

The utricle and saccule have particular orientations—the utricle lies horizontally, while the saccule is vertical. The crystals (like all rocks) naturally fall in the direction of gravity. When you move, the crystals move, too, mushing and pushing the gelatin-like membrane they are attached to, which bends the underlying hair cells, which send information about the change in position to the vestibular nerve, which transmits the information to the brain.† When your head is level and still, the position of the utricle is also flat and level, which means that its hair cells

---

*If the cupula is like a cone of gelatin, the utricle and saccule are like wider, bedazzled mounds of gelatin.

†Yes, this is the same vestibular nerve that collects information from the semicircular canals. Think of it as a mighty river with many tributaries that feeds back to an area in the brain called the vestibular nuclear complex.

aren't sending any signals; but at the same time, the saccule is vertical, and gravity pulls the crystals downward. As your head position changes, the crystals shift position, transmitting continuous information about gravity. In that way, our bodies always know which way is up.

In addition to sensing gravitational pull, the crystals in the utricle and saccule also help us sense linear motion. You know that feeling when you're in a car that unexpectedly accelerates and your head feels thrown backward for a moment? That's because your head experiences inertia. On the delicate stalk of your neck, your relatively heavy head temporarily lags behind when the car speeds forward, giving you the feeling of being pushed backward against the headrest.* The same forces work on the utricle and saccule—heavy crystals sit atop the beds of delicate hair cells. The utricle, positioned flat and horizontal, senses forward and backward acceleration in the same manner as a body in a car with surging speed. The saccule is positioned at a ninety-degree angle from the utricle, so when you are standing straight and still, the heavy crystals bend down toward your feet.† They send signals when you move vertically up or down, like when you ride in a fast elevator or take a drop on a roller coaster.

So, to summarize: the semicircular canals provide information about movement of the head based on the fluid shift inside the canals, while the utricle and saccule give the brain information about gravity, linear movements, and acceleration (direction and speed).

And just think: all that happens in an area the size of a pea.

---

*In the same way, if the car stops abruptly from, say, hitting a tree, your head will seem to be propelled forward and, unless you're wearing a seatbelt, you'll hit the windshield.
†For some reason, I imagine chocolate bars mounted on springs and stuck onto a vertical wall in a house. The chocolate bars will droop downward.

We have two such pea-sized systems, one on each side of the head as part of the inner ear. They are embedded in the temporal bones of the skull, in a location approximately behind your eyes.* They work in tandem to make sure our brains always know the position, speed, and rotation of our bodies, as well as which way is up. They make up the core of our sense of balance.

Feeling Giddy?

In 1918 Dr. Isaac H. Jones wrote:

> The stability of the earth is one of the most fundamental facts of our experience—one of our essential concepts; it is no wonder that the ancients found it hard to believe that the earth moved. Therefore, if an earthquake shakes the very ground under his feet, man is astounded and feels that even the foundations of his reason and his hope for the future are being snatched away. . . . The dizzy man carries the terror of earthquake wherever he goes, feeling that for him all order of the universe is at an end; he is disabled for to-day [sic] and disheartened for the future.

When things go wrong with the vestibular system, the earth moves—but not in a good way. The basic expectation that we stand solidly on untrembling ground is shattered in the face of an inner ear disorder. While different vestibular problems can create different clusters of symptoms, the ones most common are dizziness, vertigo, nausea, vomiting, and sometimes changes in hearing, including hearing loss or a ringing or roaring in the ears. Think severe motion sickness, with possible sound effects. In the first century AD, the Greek physician Aretaeus wrote: "If

---

*These tiny, delicate structures are surrounded by the hardest bone in the body: petrous bone. *Petrous* means stony or rocky. So, yes, in this way we are all quite hardheaded.

darkness possess the eyes, and if the head be whirled round with dizziness, and the ears ring as from the sound of rivers rolling along with a great noise, or like the wind when it roars among the sails, or like the clang of pipes or reeds, or like the rattling of a carriage, we call the affection Scotoma (or Vertigo)." A rather dramatic statement, but the sentiment reflects how overwhelming lack of balance can be.

Not all balance problems are the same or present with identical symptoms. Some disorders can make it difficult to walk at night or in low-light conditions, while others can create such a powerful feeling of disorientation that standing up is nearly impossible. In some rare conditions, such as the later stages of Ménière's disease,* you can even experience a sudden fall known as Tumarkin's otolithic crisis (known more descriptively as a drop attack), which has been described as the sensation of being pushed to the ground by invisible hands.

For many years, the terms *dizziness* and *vertigo* were interchangeable.† Most people still consider them to be synonyms. However, if you want to get picky about it, there are distinctions.

---

*Ménière's disease, or Ménière's syndrome, is a medical mystery. It affects both balance and hearing, making itself known only by the cluster of symptoms it produces—recurring episodes of severe vertigo along with hearing loss and a feeling of "fullness" in the affected ear. There is no definitive way to diagnose Ménière's disease, and there is no cure, although medications can help manage symptoms. And no one really knows what causes it. For a long time, the common theory was that the disease was caused by an accumulation of too much endolymph (that beery liquid inside the structures of the inner ear), but lots of people have excess endolymph without any symptoms at all. More recently, researchers have begun to suspect that Ménière's is an autoimmune disease. Viral infections, genetics (a family history of the disease), and migraines may also play a role. All we know for sure is that Ménière's causes extremely nasty symptoms.

†And until the early 1900s, the word *giddy* also meant the same thing. The 1901 edition of *Dr. Gunn's New Family Physician Home Book of Health* (the 230th edition!) says that "Vertigo, or Giddiness, called also Dizziness . . . consists of what is called a 'swimming in the head;' everything seems to the patient to go round; he staggers, and sometimes is in danger of falling."

Dizziness is a feeling of lightheadedness, like what you might feel if you stand up too quickly and the blood needs an extra moment to reach your brain, or that moment just before you faint. It's also similar to feeling buzzed after one or two beers—you're not drunk, just slightly floaty. Vertigo, on the other hand, is the feeling that the room is spinning around you or that you are spinning when you know that you're actually standing still. This is kind of like being very, very drunk with the "whirlies." Because the room feels topsy-turvy-spinny, it is difficult to maintain balance at all. And as if that weren't bad enough, vertigo is usually accompanied by nausea and vomiting, likely caused by the same mechanism that causes motion sickness: sensory conflict. In this case, the eyes and body know that you are standing perfectly still, while a broken vestibular system tells the brain that you are moving. The result is illness.*

Ask Kim Cathey. Kim knows.

Benign WHAT?!?

I met Kim serendipitously when I got lost on my way to visit the Department of Otolaryngology at the Duke University Medical Center in Durham, North Carolina. I found the parking garage just fine, but the path from my car to the offices was a maze of hallways, elevators, staircases, and doorways-inside-of-doorways. I must have looked stressed and confused, because Kim, who works at Duke, offered to escort me. It turns out that Kim was not just an employee. She had also been successfully treated for the most

---

*In his 1901 book, Dr. Gunn recommends treating vertigo/giddiness by making the individual vomit with a purgative or emetic and advises that "the feet should be bathed frequently, and rubbed well." Does that mean I can ask my doctor for a prescription for weekly pedicures?

common disorder of the inner ear, benign paroxysmal positional vertigo (BPPV*).

Remember those tiny ear crystals that are pulled by gravity in the utricle and saccule? Sometimes one or more of the crystals comes loose, maybe because of a blow to the head or an infection but usually just as a happenstance of aging. The crystals dislodge from the membrane, become floating debris, and eventually dissolve. But sometimes some of those ear rocks from the utricle[†] find their way into one of the semicircular canals, where they mess with your balance. Because the semicircular canals send signals to the brain about the movement and position of your head, even small bits of debris can change how positional shifts are registered. The extra weight of those tiny ear rocks alters how the fluid moves within the canal, so the vestibular system ends up sending false signals to the brain that you are moving, even when you are standing still. That means that brain signals don't match actual head movements. As with every other case of conflicting vestibular signals, that means trouble—dizziness, vertigo, nausea, vomiting, and a lack of balance.[‡]

That's what happened to my guide, Kim. Looking at her now, it is difficult to believe that her life was ever anything but serene. She seems to radiate a kind of Zen calmness, with a ready smile

---

*Experts have become so used to simply calling the disorder BPPV that even they sometimes have trouble remembering the official name. So we'll just call it BPPV, too.

†The saccule is too far away from the semicircular canals to contribute BPPV-causing crystals.

‡One interesting side note about BPPV is that what we now know to be the cause of the disorder—misplaced ear rocks in a canal—began as an assumption. Even today, with all our fancy medical equipment, there is no way to see a wayward crystal inside one of the semicircular canals. X-rays of the head only show bones of the skull. MRIs and CT scans are of no use in visualizing what happens inside those tiny loops that are protected by the hardest bones in the body. The only way that scientists were able to grasp the cause was by dissection after surgeries that required removal of the semicircular canals.

and gentle eyes. But she spent years clinging to walls and doorways for balance.

Her first symptoms came as they usually do with BPPV: first thing upon waking up in the morning. That's because when we're lying down, floating crystals tend to settle downward, usually into the posterior semicircular canal, which is the closest to the utricle. On that first morning, Kim remembers waking up, turning over, standing up . . . then falling splat onto the floor. For the rest of the day, if she turned her head or eyes, even something as mundane as glancing at her husband across the dinner table, she became so dizzy that she would have to hold on to whatever was near to keep from falling over. By the end of the day, she was thankful just to be able to lie down and go to sleep. But bed was not her friend. She held on to the sheets for balance as her body felt like it was falling in an unending spiral, Hitchcock *Vertigo* style. This disturbing sensation was accompanied by nausea.

"I got very, very, very nauseous," Kim told me. "Lying down in bed, I would have to have a trash can next to me. I would try to stay really super still and close my eyes, but I would feel the vertigo inside my head. My body was spinning, like when you're in those teacups at Disney World. I could feel myself spinning, even though I knew I was lying still."

Eventually, the motion would stop but only as long as Kim didn't move her eyes, head, or body. But, of course, no one can stay in one position all night long. Every time she turned over, the vertigo would start up again in a sickening, frightening, dangerous cycle.

And that was just day one for Kim. She suffered for years. Years! Each episode could last for days or even weeks before it would go away for a time, but eventually another bout would hit her. I asked if she had seen a doctor, and she had. Several of them, many times over, including a neurologist. She begged for some kind of relief. As Kim said, "I kept telling my doctor: Look, this is driving me crazy. I can't get down on the floor, I can't lower my head. I can't turn my head. I feel like I'm falling. I'm sick as a dog."

But no one could offer any real help. The doctors Kim saw thought she was merely stressed, or maybe the symptoms were due to her high blood pressure, or perhaps she was standing up too quickly. She left each office visit feeling frustrated, like she wasn't being heard. This wasn't just "standing up too quickly." Kim's life and head were out of control. She was a mass of bruises—on her shoulder from when she hit the wall, on her knees from when she fell to the ground, on her legs from when she tripped and smacked into the stairs. During the worst of it, she couldn't walk at all and had to crawl. The symptoms were totally debilitating. Between episodes, Kim still couldn't enjoy life because she was paranoid, wondering when the next bout of spiraling, nauseating imbalance would hit.

This went on for three years.

Enter Dr. Richard Clendaniel, assistant professor at Duke University and the guy who wrote the book on vestibular rehabilitation.* He is also a physical therapist and Kim's superhero, although she didn't have much hope when she first met him.

"I thought, there's no way this guy's going to be able to help me," Kim remembered. "It was a regular physical therapy practice. There was a guy on one side of the room doing leg exercises after knee surgery, and a woman on a machine to rotate her shoulder on the other. But I was desperate. By that time, I thought I was losing my mind."

---

*That's figurative. In reality, he coedited and wrote chapters for the book *Vestibular Rehabilitation*, fourth edition (Philadelphia: F.A. Davis, 2014). Although Dr. Clendaniel is modest about his standing in the community of balance professionals and would never say this about himself, he is one of the top researchers and clinicians in the field.

I must admit that when I visited Dr. Clendaniel in his office at Duke University, I had expectations of being wowed by gadgets, machinery, electrodes, lights, buzzers—you know, science fiction movie devices. Nope. There was nothing like that. When you're really good at what you do, you don't need the bells and whistles.

"We're what you might call low-tech," Dr. Clendaniel said.

I would call it low-key in a good way, and Dr. Clendaniel himself is a reflection of that philosophy. His body language is fluid and relaxed, fitting for someone who has spent his life studying human movement. Most importantly, Dr. Clendaniel is soft-spoken, with the kind of voice that could narrate relaxation meditations. Smooth and soothing. After a hectic morning's drive, I could feel my body calming down during our conversation. He listened intently, then provided a considered response to my questions. He also had an answer for Kim's problem.

During Kim's office visit, Dr. Clendaniel put a black box on her head and told her to move her eyes in different directions. The box was actually a pair of goggles that could block out light and distractions while magnifying the eyes for an observer.* Inside the goggles, a tiny camera focused on Kim's eyes, capturing every movement. Outside the goggles, Dr. Clendaniel could watch Kim's eyes on a monitor, looking for sharp, rattling, quivering, or rotating motions when she looked in a specific direction. Kim remembers thinking that it all felt like so much hocus-pocus. But within a few minutes, Dr. Clendaniel had figured out that Kim had BPPV. Although Kim's experience was on the more severe end of the vertigo/dizziness symptom spectrum, what she felt was typical for people with the disorder. Dr. Clendaniel explained what was happening, and then—miracle!—made it all go away in minutes.

---

*I tried the goggles on. Physically they feel a lot like a diver's mask.

"I about started crying," Kim said. "I said, 'It's gone!' I got up and I grabbed him and hugged his neck and kissed his cheek. I don't think he was expecting that."

What's surprising to me is that it took so long for someone— *anyone*—to recognize the common symptoms of BPPV and to offer help. BPPV is the most common cause of dizziness and vertigo, accounting for about 20 percent of all vertigo cases and affecting more than four million people in the United States each year. It could be that a decent treatment for BPPV wasn't developed until the 1980s and was not widely recognized until the 1990s. Until then, people suffering with BPPV basically had three options: (1) do nothing, taking comfort in the knowledge that a diagnosis of BPPV was not life-threatening or a sign of stroke or brain tumor;* (2) take sedating medications to get through the day until the symptoms went away; or (3) undergo surgery to cut one of the ampullary nerves, which feed signals from the semicircular canals to the vestibular nerve, and then on to the brain. While the surgery worked really well at eliminating symptoms of BPPV, one of the possible side effects was severe hearing loss.

But in the 1980s, Dr. John Epley devised a simple but highly effective treatment for BPPV, which uses a series of head movements to move the tiny crystals out of the semicircular canal they are stuck in. What's used today, and what Dr. Clendaniel did for Kim, is called a modified Epley maneuver.

OK, I'm going to give a brief description of the maneuver to give you an idea of how simple it is. However, experts I've spoken with, including Dr. Clendaniel, recommend that you see a physician or physical therapist before attempting to use this exercise

---

*In 1982, respected neuro-otologist Robert W. Baloh wrote: "A simple explanation of the nature of the disorder and its good prognosis gives a patient a great deal of relief."

to treat yourself, for a number of reasons. First, if you are having troublesome vertigo symptoms, you'll want to make certain that your diagnosis is actually BPPV and not something else, possibly something more serious. Second, this particular maneuver treats crystal displacement into the posterior semicircular canal. If you have crystals in one of the other two canals, these exercises could potentially lodge them deeper inside. And finally, you'll want to let your health care provider show you the correct way to do the maneuver so that it is most effective. There are some things that cannot be described in print. For example, one of the first movements involves turning your head forty-five degrees toward the affected side. Do you know what turning your head forty-five degrees feels like? And which is your personal affected side? Let a skilled clinician show you the process once, and then you'll know it forever.

Sitting up on the side of your bed, turn your head forty-five degrees to the affected side (the side that seems to make you feel the worst). With your head still in the forty-five-degree turn, lie down quickly. (Some doctors recommend hanging your head over the edge of the bed, but others say to lie flat.) This will make you feel really sick, but don't worry—the sensation goes away within about fifteen seconds. Stay in that position for thirty to sixty seconds, then turn your head so that it is forty-five degrees to the opposite side. Wait another thirty to sixty seconds. Then turn your body and head together in the same direction so that you are lying on your side with your head pointing down toward the floor at a forty-five-degree angle. (To do this last move, pretend that you have a stiff neck that can't move—when you turn your body, your head moves with your body, without independent movement from your neck.) Stay that way for another thirty to sixty seconds. Finally, keeping your head turned, sit back up in the original position. You can look forward again, but keep your head level and sit still for about ten minutes.

If you imagine the floating crystals in your ear, you can see that this maneuver is about moving the head in a way that encourages the rocks to move back to the opening of the semicircular canal, and then down and away from the opening. The modified Epley maneuver* isn't a permanent cure, but it can be done at home anytime the vertigo hits. Some people need to repeat this exercise for two or three days before they find relief. Most others, like Kim, lose their vertigo immediately. Kim still has episodes, and the vertigo and nausea are just as intense. But now she can lie on her bed, do the repositioning movements, and get on with her life.

I love a happy ending.

Except, wait . . . I missed something. How do eye movements help doctors diagnose BPPV? What do the eyes have to do with minute grains of crystal misplaced inside the loops of the semicircular canals in the ear?

---

*It is "modified" because the original maneuver also used vibrational instruments against parts of the skull to help rattle the crystal out of its annoying position.

3

The Eyes Are the Windows to the Ears

Vision

"**S**O, DID MY EYES LOOK 'NORMAL'?"
I have just experienced the rotary chair, one of the diagnostic tools used to look for problems with the vestibular portion of the inner ear. Actually the test ended about twenty minutes ago, but I have been too afraid to ask what my results look like. I climbed into the rotary chair strictly for research; it was not part of a medical examination, and I don't feel at all off balance, so there really is no reason for me to worry. And yet, who knows what my eyes are saying? Is there a hidden problem I might not be aware of that revealed itself in the dark?

I laugh. Even to my own ears, I sound nervous.

This test began with wanting to understand more about the relationship between the eyes and the vestibular system. How can doctors diagnose benign paroxysmal postural vertigo (BPPV) simply by looking at a patient's eyes? To that end, I visited Dr. Erin Piker at her offices* at the Duke University Medical Center Department of Otolaryngology–Head and Neck Surgery in Durham, North Carolina. Dr. Piker holds both a doctor of audiology degree (AuD) and a PhD, and she is strikingly lovely. It's my own prejudice that I expect researchers to look a little geeky or at least to be cloaked in a lab coat. But Dr. Piker is wearing a long, silky patchwork dress, with blonde hair tied up, fashionably messy. And she is whip-smart. She answers every question as quickly as if she were on a game show, but without a contestant's uncertainty.

Dr. Piker explains that some eye movements are yoked to signals from the semicircular canals through something called the vestibulo-ocular reflex (VOR), a hardwiring of the motion of the eyes to movements of the head. Its purpose is to keep vision stable, regardless of how you move. Without the VOR, the world would be an out-of-focus, blurry mess, and it would be impossible to maintain positional balance.

Check out how this works: Extend an arm straight out from your body, and hold up one finger.† Focus your eyes at the tip of your finger. Now shake your head side to side, as though you're saying no. Do it again, faster, always keeping your eyes on the tip of your raised finger. No matter how fast you shake your head, the tip of your finger stays in focus. Now do the exercise again, but this time move your head in slow motion. Notice how your eyes move to stay trained on your fingertip. If you turn your head far to the left, your eyes (still focused on your fingertip)

---

*Since my visit, Dr. Piker has moved. She is now assistant professor and director of the Vestibular Sciences Laboratory in the Department of Communication Sciences and Disorders at James Madison University in Harrisonburg, Virginia.
†The choice of finger is up to you, depending on your mood.

will move all the way to the right. And if you turn your head far to the right, your eyes will move all the way to the left. This may seem like a lame demonstration. I mean, of course if you look at your finger your eyes will move. But there's something deeper at play here. For contrast, raise your finger again, but this time, as you focus on your fingertip, hold your head still and move your finger back and forth quickly. Your finger will look out of focus and blurry.

The difference is head movement. When you move your head, the environment (your fingertip) stays in focus; when the *environment* moves, focus is lost. As you move through the world, your eyes move instinctively and automatically to keep the world stable and in focus. You don't have to tell your eyes to do that; they are wired to the vestibular system through an area deep in the brain called the vestibular nuclear complex in such a way that your eyeballs automatically move to compensate for your body's movement. When you shake your head side to side, the thick endolymph fluid in the horizontal semicircular canals (one on each side of your head) gets displaced, moving the tiny hair cells at the end of the loops, which send signals about your head position to the brain. The brain reacts and sends simultaneous signals to the eyes, telling them to move in a manner and speed that is exactly proportional to the head movements.

It's a reflex. It's fast, automatic. Think about another reflex that's easier to see, and one you have probably experienced at least once during a visit to the doctor: the patellar reflex (also called the knee-jerk reflex). Sit straight in a chair and cross your legs at the knee, making sure the top foot dangles freely over the bottom leg. Have someone give a gentle karate-type chop to the area just below your kneecap. This tap on your leg activates the patellar tendon, causing your lower leg to kick out. It's an automatic reflex, not something you consciously control. The VOR has the same immediate quality: your head moves,* and your eyes move in instantaneous, reflexive response. That involuntary eye movement is called nystagmus.

Because of that strong and permanent connection, it is possible to diagnose some inner ear disorders by simply watching how the eyes move after a variety of diagnostic challenges. And that's what I hope to learn when I visit Dr. Piker at Duke University Medical Center.

Watch the Bouncing Eyeballs

Room 1 of Dr. Piker's domain is dominated by the equipment for videonystagmography (VNG), which tests for inner ear function by measuring how the eyes move in response to a particular stimulus. It's not possible to see the inner ear from the outside. Even using one of those ear lights† shining down through the ear canal, the farthest you can see is the eardrum, which is part of the middle ear. The eardrum, like a membranous door, blocks us from seeing anything deeper inside. We would have to break through the eardrum or drill through parts of the skull to get a real look. But the eyes are right up front, and thanks to the VOR, diagnosticians can see problems of the inner ear reflected in eye movements.

Take BPPV,‡ which causes dizziness and vertigo when tiny ear rocks (otoliths) become dislodged from the gravity-sensing, crystal-laden utricle and make their way into one of the semicircular canals.

---

*Of course, the head is quite often connected to a body. The VOR is not merely about keeping eyes focused as you move your head back and forth; I say "head" because the nerve connections for this reflex are all in the head. The VOR functions anytime you—your head and your body—move around in the world.

†Officially, it is known as an otoscope.

‡This disorder is described in more detail in the previous chapter.

When a wandering otolith gets lost inside one of the semicircular canals, it displaces the thick endolymph fluid and causes tiny hair cells to bend, which sends a signal to the brain that mimics the signal that says, "The head is moving now!" That signal is the same as the one sent when the head actually *is* moving. The hardwired VOR causes the eyes to move spontaneously in a way that would maintain visual stability *if the head had actually been moving.* That happens whether the head movement is real, like when you focused on your fingertip and shook your head, or the result of a false signal—which happens when an ear crystal becomes lodged in one of the semicircular canals.

The exact type of nystagmus triggered by an erroneous "I'm moving" signal depends on which semicircular canal contains the wandering crystal. Eyes can shoot back and forth, up and down, or sometimes in a circle. That sounds dramatic, but if you have nystagmus from a vestibular disorder, you won't necessarily know it. The eye movements aren't constant, and they can be subtle, just a slight quivering or jerking motion.* The worst symptom is usually balance-related: a feeling of motion sickness with vertigo. Your vision may not change measurably, either, although it may be more difficult to focus on an object, just as it was difficult to focus on your fingertip when you shook it back and forth. But to a vestibular expert like Dr. Piker, nystagmus is like a giant, neon-lit arrow pointing to the source of the balance symptoms. With the help of VNG, she can pinpoint the source of a person's dizziness by measuring, magnifying, and recording even the smallest eye movements. And the main portion of VNG is the caloric reflex

---

*In some cases, nystagmus is very noticeable. Eye movements can be really large in some people or with some medical issues. Severe concussion can potentially cause the eyes to shake or roll in a way that is obvious to any observer. And some people have infantile nystagmus, which develops early in life and is often more pronounced.

test.* It is still considered the gold standard of inner ear testing, even though the basic procedure was pioneered more than one hundred years ago by Dr. Róbert Bárány.

Hot Water, Cold Water, Just Right

By all clues, Dr. Bárány was a very serious individual. Exhibit One: any of his official photos. Granted these were taken in the early 1900s when smiling for posterity just wasn't done, but his portraits make him look like a slightly deranged stage magician who just might hypnotize you into turning over your bank account numbers. Either that or a suited-up hipster who drank too much coffee. Exhibit Two: his personal history. Bárány's intensity was likely forged from the physical pain of having tuberculosis of the bones when he was a child, an illness that left him stiff-legged throughout his life. He was also brilliant, graduating from the University of Vienna when he was just eighteen years old and completing medical school six years later, in 1900. He studied under Sigmund Freud, a man Bárány greatly admired, although the admiration seems to have been one-sided.[†]

That combination of intensity and intelligence set Bárány up for greatness. He was awarded the Nobel Prize in Physiology or Medicine in 1914 for his discovery that involuntary eye movement (nystagmus) was a reflex action of the semicircular canals.[‡] With

---

*A calorie is a unit of heat energy. While most of us use the word *calorie* willy-nilly when talking about that mysterious quality of food that makes us fat if we eat too many of them, the nutritional meaning is the same. A calorie is the amount of energy it takes to raise one gram of water by one degree Celsius. The caloric test uses a change in heat or temperature. Sadly it will not contribute to weight loss.

†Freud claimed to have dismissed Bárány as a pupil because "he seemed to be too abnormal." No additional information was given, nor support made, for this rather vicious remark.

‡He was unable to receive the honor in person, however, because at the time the prize was announced, he was a prisoner of war, after volunteering his services as a civilian surgeon

the help of a particularly astute (yet anonymous) patient, Bárány discovered that it was possible to diagnose a dysfunction of the semicircular canals simply by irrigating the ears with hot or cold water. When water slightly warmer than body temperature circulates through the middle ear area, within seconds the heat permeates the inner ear space, where it heats up the endolymph fluid. The heat changes the density of the fluid, causing it to move a little through the semicircular canals and bend the tiny hair cells, which send signals to the brain that say, "Hey, the head is moving!"* The head isn't moving, of course—the process was tricked by the heat. But even fake head movements can trigger the VOR, resulting in the automatic eye movements of nystagmus.

The discovery of this connection between temperature change and nystagmus was pure happenstance. In his acceptance speech for the Nobel Prize,† Bárány related his eureka moment:

> One of my patients, whose ears I was syringing, said to me: "Doctor, I only get giddy‡ when the water is not warm enough. When I do my own ears at home and use warm enough water I never get giddy." I then called the nurse and asked her to get me warmer water for the syringe. She maintained that it was already warm enough. I replied that if the patient found it too cold we should conform to his wish.§ The next time she brought me very hot water in the bowl.¶ When I syringed the patient's

---

to the Austrian army in World War I. He wasn't released until 1916, and then only by personal intervention of Prince Carl of Sweden. So I think Bárány can be forgiven for looking a bit keyed-up in his photos.

*It's the same false signaling that happens when a roaming ear crystal floats into one of the semicircular canals.

†It was delivered belatedly because of the whole prisoner-of-war thing.

‡As discussed in the previous chapter, *giddy* used to refer to feelings of dizziness.

§Snarky nurse almost derailed the whole discovery!

¶Is it just me, or does the nurse's behavior strike you as a bit passive-aggressive?

ear he shouted: "But Doctor, this water is much too hot and now I am giddy again." I quickly observed his eyes and noticed that the nystagmus was in an exactly opposite direction from the previous one when cold water had been used. It came to me then in a flash that obviously the temperature of the water was responsible for the nystagmus.

Using this method, Bárány went on to describe and analyze the phenomenon we now know as BPPV (which had previously been called positional vertigo of Bárány). Doctors still use the hot water–cold water caloric reflex test to diagnose or confirm a problem within one of the semicircular canals. From what I understand, it's not a pleasant experience. The operational part isn't bad—you start by lying down on your back with your head on a pillow. Then the tester floods the ear with slightly warm or slightly cold water* via what looks like a small hand drill with a Waterpik-like tube in place of a drill bit. The water goes in, hits the eardrum, and comes right back out in a continuous flow for about thirty seconds. It feels like your ear is full of water (because it is), and there's a loud *zzuh-zzuh-zzuh* sound, a combination from the machine motor and water pressure. The unpleasantness comes from the outcome of the test, when the VOR is triggered and you feel like the room is spinning.

Dr. Piker can target the fluid to affect the endolymph in the horizontal semicircular canal by having a patient lie on his or her back during the test. Then, when the VOR is triggered, it causes horizontal eye movements—the quick, jerking nystagmus moving left and right (versus up and down or in a circular pattern). By putting water in one ear at a time, Dr. Piker can differentiate between

---

*Sometimes hot or cold air is used instead of water. The outcome is the same, regardless of whether water or air is used.

the functions of the left and right inner ears, independent from one another.* What you want is a strong, symmetrical response that shows both sides of the vestibular system performing equally well. If the results are asymmetrical, with different motions and responses, that points to a clear problem on one side. In fact, the caloric reflex test is currently the only method health professionals have for diagnosing unilateral problems in the vestibular system.

In Dr. Bárány's day, the only eye measurements he took were the ones he could see; he would irrigate the ear, then watch for the telltale bouncing eyeballs. Today, physicians and researchers have video goggles. When you put them on, your eyes are magnified, enhanced, and displayed on a video screen, and a mathematical readout charts eye movements (it looks similar to the heartbeat tracings from an EKG). So if you have an episode of nystagmus triggered by the caloric test, Dr. Piker has a magnified view of your eye movements and a waveform graphic display. Unfortunately all you'll see is the world spinning around from vertigo.†

Dr. Piker puts the goggles on so I can see what she sees when a patient is in the chair. The goggles themselves are huge. They come down as far as the end of her nose, and they stick out a good five inches or so from her eyes. At the top of the goggles is another thick bar that looks like a surge strip with two cords plugged into

---

*Interestingly, whether you use heat or cold makes a big difference in how the eyes move. That's because heat and cold create fluid density changes inside the inner ear that reverse the normal flow inside the semicircular canals. Experts remember the difference by the mnemonic COWS: Cold-Opposite Warm-Same. The horizontal eye movements elicited by the caloric test have different speeds as they move right or left; they are not steady and equal like a metronome. There is a fast beat to one direction, followed by a slow return. COWS tells us that if cold water is flushed into the left ear, it will have a fast beat to the right (i.e., the side opposite from where the water is), followed by a slow drift back left; and if warm water is flushed into the left ear, there will be a fast beat to the left (i.e., the same side as the water).

†If you need to get this test, rest assured that the triggered episode of vertigo is short, usually no more than a few minutes.

it. On the computer, I can see a video display of her eyes and an electrical readout that corresponds to her eye movements. Then Dr. Piker changes the display to show a video of just one eye—one *enormous* eye that fills the entire computer screen. For a professional, the electrical waveforms give the most accurate information about eye movement. For the average nonprofessional, however, the single-eye video shows nystagmus very clearly. With the video of an eye blown up to nearly two feet wide, I can see every tiny twitch, jerk, or scan of the eyeball. It's mesmerizing. We watch videos from six different people undergoing the caloric test, and by the end I feel confident that I could diagnose a problem with the semicircular canals just by watching eyes.*

A newer diagnostic tool is the vestibular evoked myogenic potentials (VEMP) test. This test is especially cool because it gives doctors access to a part of the inner ear they couldn't tap into before—the gravity sensors—all by recording a surface muscle reflex. Unlike the caloric test, the VEMP doesn't evoke dizziness, and it doesn't cause any additional vertigo or nausea for the already-sick patient. For the VEMP, electrodes are placed on the skin over the sterno-cleidomastoid muscles in the neck† or on the muscles surrounding the eye.‡ As odd as it may seem, those surface muscles have reflex connections to the saccule and utricle gravity sensors deep in the inner ear. The saccule has connections to the sternocleidomas-toid muscles in the neck, and the utricle is linked to the muscles around the eyes. During testing, short bursts of sound are directed

---

*In reality, I probably can't—and won't. But I learned enough to be able to recognize nystagmus and make a recommendation to see a health care provider. Of course I fully expect my friends to avoid direct eye contact with me after they read this chapter.
†Place your right hand lightly on the left side of your neck, then turn your head all the way to the right. The large band of muscle that pops out under your hand is the sternocleido-mastoid. There is one on each side of your neck.
‡Specifically the inferior extraocular muscles.

to the inner ear,* stimulating the gravity sensors. The electrodes track and record muscle reflexes in response to the sound waves. In the case of the saccule, the doctor directs sound waves to the inner ear, while the person being tested is asked to turn her head and flex the sternocleidomastoid muscle. With sound stimulation, the inner ear temporarily inhibits part of the surface muscle, and the VEMP picks up that change and translates it to an electrical waveform. By examining that little waveform tracing, experts can read volumes in terms of how big the muscle response is, the size of the response, and how long it takes for the response to show up. Ideally you want both sides to have symmetrical readings, which would suggest normal function of the saccule in both ears. To test the utricle, the same sound-pressure test is given, but this time the electrodes are placed around the eyes.

That's a complex description. Here's what's important: The VEMP test is remarkable because it is the only way to get an indirect glimpse of the functioning of the gravity sensors. In addition, it highlights another balance revelation—the muscles of the neck play a role in balance and dizziness. I hadn't expected that; it didn't dawn on me that it could be true until Dr. Piker explained the VEMP. But the neck-balance relationship affects a lot of people. In the 1980s scientists began talking about cervicogenic vertigo,† debating whether it was a real condition. Thirty years later, the debate continues. There are no tests that can pinpoint a diagnosis of cervicogenic vertigo. It is only diagnosed by a process of elimination. If you were to complain of vertigo that began at the same time as neck pain, you wouldn't get a diagnosis of cervicogenic vertigo

---

*Totally painless.
†*Cervico* means pertaining to the neck; *genic* means a beginning. So *cervicogenic vertigo* (also sometimes referred to as cervicogenic dizziness) means vertigo that begins in the neck.

until all other possible diagnoses are eliminated. Even then, some physicians hold the position that the condition must be caused by some other linking factor, such as mild brain injury. Still, there is an abundance of literature and anecdotal reports from clinics about vertigo and/or dizziness that began after a whiplash injury, chronic neck muscle spasm, or athletic injury involving the neck (such as injury from heading the ball in soccer, or a helmet hit or just about any other injury from playing football).* The VEMP test lends credence to this relatively new diagnosis.†

That Looks Really Sadistic

Dr. Piker's second diagnostic room is, at first glance, frightening. It is dominated by a huge, black cylinder with an open door. Inside is a . . . chair? OK, let's call it a chair. It is, in fact, a rotary chair. But calling it a chair doesn't begin to describe how intimidating it looks. The rotary chair is hiked up on what look like hydraulics and rotors; it has a large overhead extension with a laser target generator, digital eye tracking system, microphones, and speakers. The sitting area is well padded, with the sides of the seat turned up to keep you in place, an abundance of belts to strap you in, and large controller shafts at the end of each armrest. It is a highly sophisticated vestibular and neuro-otologic testing station. It looks like a torture device.

My first words upon entering the room are: "That looks really . . ."

---

*Physicians at the Cleveland Clinic have even reported a swimmer who had short intervals of "room-spinning vertigo" whenever she swam freestyle, which requires repetitive neck turns.

†Because cervicogenic vertigo is still controversial, some physicians won't make that diagnosis. A person with vertigo or dizziness along with neck pain may need to visit a dedicated balance center to get a diagnosis and appropriate treatment.

"Creepy?" Dr. Piker offers.

I was thinking *sadistic*, but *creepy* works, too. I'm not surprised to learn that the rotary chair was pioneered by Mr. Intensity himself, Róbert Bárány, who designed and constructed spinning chairs especially for treatment of vertigo with nystagmus. One noted physician of the time, Dr. Isaac H. Jones,* wrote:

> In order to carry out the turning-test with any degree of accuracy, it is essential to have a revolving-chair especially constructed for this purpose. Not only must it revolve smoothly, but it must have the necessary attachments for holding the head in the proper position as well as a stopping device to clamp it instantly and firmly. Barany has constructed a chair for this purpose. He sent us one of these chairs . . . which we were fortunate enough to receive only a few days before the outbreak of the war in Europe. So far as we know it is the only one in this country.

Dr. Piker's modern-day rotating chair is enclosed in a black miniroom to prevent visual distractions. She adjusts the fit of a pair of giant black goggles, which cover most of my face, then straps me into the creepy chair. It is disconcertingly comfortable. I wouldn't mind at all if I had to use this as an office chair. Dr. Piker leaves the cylinder and closes the door. It is bottomless-pit dark.

The first step is to calibrate the machine for my eye movements. For the patient (me), that requires nothing more than sitting in the dark and looking at a red dot projected in front of me. The calibration here is especially important because the device has to capture the movement of my eyeballs while the chair I'm sitting

---

*Apparently there was great interest in the rotary chair because it was so easy to administer—none of that pesky water irrigation and no need for the patient to follow complex instructions. In fact Dr. Isaac H. Jones instructs other physicians that "this is very useful in stupid individuals, or in patients who are weak or irritable."

in rotates in a circle. The goggles have a mirror that reflects an image of the eyes to a camera overhead, which then relays the visual display and data points to Dr. Piker's computer.

Remember how the water irrigation made the eyes move by creating false signals of head movement? Well, with this test the movement is real. The device measures the responses of the same parts of the inner ear as the caloric test—the semicircular canals—but the rotary chair looks at the functioning of both ears simultaneously. This test is best for diagnosing bilateral inner ear problems, which could be caused, for example, by taking ototoxic medications, which can cause permanent damage to portions of the inner ear.* The main test is a slow rotation, what's called a sinusoidal rotation, which tests vestibular function at different frequency ranges. The test is repeated several times in a row, each time with a different frequency, which requires different rotational speeds. Some are so slow you barely know the chair is moving. Dr. Piker can also ramp up the speed so that you'll be grateful for all the straps and padding on the chair. If the chair is turning to the right and you are asked to keep your eye on a spot on the wall, your eyes will have to turn left, the direction opposite from the movement. That's the same principle demonstrated at the beginning of this chapter, when you held your finger still and shook your head back and forth. While I am spinning in the rotary chair, Dr. Piker is at her computer looking for the timing of my eye movements, how much movement there is, how symmetrical my eye movements are if the chair moves right versus left, and other bits of data.

And so I rotate. It's not a big spin, nothing crazy. The most disorienting thing is being totally in the dark. Even at night, if you turn

---

*The most notorious ototoxic medications are aminoglycoside antibiotics, including gentamicin and streptomycin, and platinum-based chemotherapy agents.

off all the lights in your house, there will probably still be streetlights, lights from your neighbor's house, and even light from the moon and stars. I can't remember another time when I was totally in the dark with my eyes open. Dr. Piker gives me the task of talking—about anything—and if I run out of things to say, she says she will ask me questions.* Chatting is important for two reasons. First, it distracts the patient from the chair's movement. Rarely does someone become sick on the rotary chair, but talking about other things can make the test easier to tolerate. Second, talking keeps the patient alert, and the more alert you are, the bigger the nystagmic reflex.[†]

Once the calibration is over, the test begins. All at once, the interior wall of the cylinder I'm in is covered in dots of white light. They start to rotate around me in one direction for about ten seconds; then they change direction. It looks like a disco planetarium.[‡] As they move, there's a little bit of vection going on, much like what I experienced in the Vominator; it feels like my chair is moving, but it's not—the lights on the wall are moving.

Dr. Piker talks to me through the speaker, letting me know that the chair will begin moving and that she'll keep it at one of the comfortable middle frequencies. My job, once again, is to look forward and keep talking. As the chair begins rotating, it has the familiar feel of one of those visual attractions at Disney World, the kind where the car you're in moves through a tunnel before entering a world of dinosaurs or ghosts. It is not at all unpleasant. Dr. Piker says that the vast majority of people she sees find this test very easy.

---

*Chatting continuously is one of the skills I have learned since moving to the South. I didn't need any questions to keep me busy.

†So, talking about my dogs and cats is not always an annoyance! Here it could help Dr. Piker to be more accurate when making a medical diagnosis.

‡Not that such a thing exists. It's just the first thought that popped into my mind. It still feels accurate.

"I've been working clinically since 2007," she says, "and I have never had anyone get sick in the chair. It's really, really easy. I have had people get sick during the caloric. That does cause a sensation much, much stronger than this. Not too many people get sick [in the chair], but maybe once every two months or so. They are the same people who would come in and say, 'I can't sit in the back of a car without getting sick.' They are sensitive to movement."

The rotary chair isn't used on all patients. The caloric test—which tests each inner ear separately and independently—is the most sensitive, Dr. Piker's "step one." The rotary chair looks at both ears together. If the caloric test shows that both ears have very low vestibular function, the rotary chair can either confirm the bilateral weakness or suggest that there was a glitch in the caloric test.* If the results from the caloric test are questionable, or if there is a big asymmetry in the vestibular functions, then the rotary chair can fill in some information about how well a person has compensated for some vestibular loss. If one side isn't working well, hopefully the other ear—along with the brainstem and cerebellum—will take over for the weaker ear, compensating for the one-side deficit.

When central compensation happens, it occurs first at a static position—that little spontaneous eye movement (nystagmus) goes away. For example, let's say you get neuritis, an inflammation of the vestibular nerve, and your vestibular function drops. One thing that can happen is that the faulty nerve sends continual signals to the brain that you are turning in circles, over and over and over again, even when you are standing still. With all those spinning

---

*Because the caloric is hard to tolerate, the test results aren't always "clean." It is impossible to read eye movements if your eyelids are shut tight while you're getting sick.

signals, you'll get sick to your stomach, with dizziness, vertigo, and lack of balance—just like what would happen if you were really physically spinning. Over time, that ear might start functioning properly again, or it could remain permanently damaged, complete with an ongoing, never-ending "spin cycle."

Eventually the brain will figure out that the spin signal is wrong, that you're not actually moving. Then your brain recalculates the information it gets from the ears, applying its own translation algorithm as a way to override or compensate for the faulty signals. Once that central compensation happens, the spontaneous movement of the eyes, the nystagmus of spinning, will stop. But that's the simple part . . . the standing-still part.

The next step is to develop dynamic compensation. When you move your head with a "broken" or dysfunctional inner ear, your brain gets mismatched signals from the two ears—one provides correct information, and the other provides false information. That's why head movements can make people with one-sided inner ear problems feel extremely off balance. Just as with the static or standing-still recalibration, your brain has to learn how to recalculate signals from your moving head, allowing your eyes (hardwired to the inner ear) to catch up and move appropriately. With the rotary chair, Dr. Piker can see how well the brain has compensated for any inner ear deficit. Once the brain compensates, the eyes will get back to moving in direct connection to head movements, the way they are supposed to move. Once that happens, you'll be able to turn your head without getting dizzy. There is hope of a normal life.

So the overall lesson is this: While the vestibular system of the inner ear is central to our sense of balance, our eyes are a crucial component as well. A kind of really important second banana. Dr. Watson to Sherlock Holmes. Shirley to Laverne. Hoda to Kathie

Lee.* The eyes and vestibular system are intertwined and function best as team. Together they perform an essential role in helping us stay balanced.

After about an hour of visiting with Dr. Piker, I am ready to leave her lab. I have a small wave of concern: She never told me what she saw while my eyes were monitored in the rotary chair. I get motion sickness easily, so I would be a prime candidate for some sort of balance disorder. Why hasn't she said anything about it? Is it bad news, and she's holding back because I'm not really a patient?

I take a breath and ask the question I've been holding on to for nearly an hour.

"So, did my eyes look 'normal'?" I ask.

"Very normal," Dr. Piker says. "So far, so good."

The warm feelings of being called normal stay with me for the better part of a week.†

---

*I take it back—Hoda is second to none!

†Yes, I know that *normal* refers only to my involuntary eye movements, but I'll take what I can get.

4

Do You Know
Where Your Body Is?

Proprioception

IN 1971 BRITISH BUTCHER Ian Waterman was as tall, healthy, and handsome as any nineteen-year-old could hope to be. Then, suddenly, his entire body went to sleep. You know that feeling when you lean heavily on your arm and temporarily lose all sensation so it feels like you have a foreign chunk of meat dangling from your shoulder? That. That's what happened to everything below Ian's neck. Except Ian's body never woke up.

We all have long nerves that run from distant parts of the body, such as fingers and toes, back to the spinal cord, where signals are consolidated and sent on to the brain. If you look at a magnified cross section of one of these nerves, you see that each nerve is actually a bundle composed of many smaller nerves, and each bundle contains hundreds of individual nerve fibers. The smallest of the nerve fibers transmits information about pain and temperature.

Other, larger nerve fibers relay information about the sensation of touch, as well as feedback from our muscles and joints, a sense called proprioception.

Of all our senses, proprioception is the most elemental—it tells us where our body parts are and what they are doing moment by moment. Here's a very simple demonstration: With your eyes closed, raise one arm above your head. Your sense of proprioception tells you exactly where your arm is—the height of your shoulder, the precise bend of your elbow, whether your fingers are outstretched or curled in. You can know all those minute details without looking at your arm because your brain continuously monitors incoming signals from your sensory and proprioceptive nerves so that you always know what your arm is doing and how it is positioned. The sensation of an arm or foot "falling asleep" happens when you compress sensory and proprioceptive nerves—you lean the wrong way for too long, applying pressure on the nerve, and nerve signals get pinched off, the same way you can fold a hose and stop the flow of water. The pins-and-needles sensation you get when your limb starts to "wake up" is just the nerve becoming active again.*

Knowing where your limbs are and what they are doing is so basic that we barely think of it. When the system fails, as it did for Ian Waterman, the results are devastating. Ian's problems began after a flu-like illness. Doctors now believe that during his body's fight against the invading virus, Ian's own immune system turned against him and attacked not only the virus but his proprioceptive

---

*When I was a child, it was common playground knowledge that a foot fell asleep because the blood supply was cut off. My friends and I were all convinced that if your foot fell asleep for too long, it would fall off. We never thought to check with an adult because, well, we *knew* it was true. That fear caused this nervous-Nellie kid quite a bit of anxiety. Apparently this is a common misconception. I just hope no one else spent as much of their childhood working so hard to avoid gangrene as I did.

nerve fibers as well. The destruction happened quickly. One day he was carving pork chops and dreaming of owning his own butcher shop, and a week later he was lying helpless in a hospital bed. Ian could see his body, but he had no sensation of it. With his eyes closed, Ian felt as though the sum total of "him" was little more than a head on a pillow.* He had no conscious sense of arms, legs, chest, stomach, or any other part of what typically makes up the feeling of self. His damaged nerves severed him not only from his body sense but also from the outside world—he had no tactile ability. He could neither feel what he touched nor know if anything was touching him. For all he could tell, his body was simply floating, with no bed beneath him.

While an autoimmune reaction destroyed just the nerves specific to proprioception and the sense of touch, Ian's other nerves survived. This means, remarkably, that Ian is not paralyzed—the nerves that allow his muscles to move are intact, as are the tiny nerve fibers that allow him to feel pain and temperature. His joints work. His muscles work. But he has no automatic, unthinking control over them. Because the body feedback nerve fibers were destroyed, his brain reacts as though his body doesn't exist. And, at least initially, his brain could not control anything his body did. For months after the viral devastation, Ian's arms and legs moved in random, almost flailing motions, as if he were a marionette with an untrained puppeteer pulling his strings. He could see his limbs, but his brain couldn't find them (neurologically) and therefore couldn't get through to tell them what to do. It would be like trying to call a friend when you don't know her new phone number—you have a phone, she

---

*Because he retained proprioception from the neck up, Ian could feel his head and where it rested.

has a phone, so communication should be possible, except you can't actually connect. Ian Waterman's body had permanently hung up on his brain.

Along with the vestibular system and vision, proprioception is one of the primary factors that contribute to our sense of balance. Science has known about that triad for a long time, but until Ian Waterman came along, scientists didn't fully understand what total loss of proprioception—*just* proprioception—meant. You can easily imagine what the world is like without the sense of sight—just close your eyes. And you can upend the function of the inner ear by going into environments where there isn't any gravity, such as in space or on a microgravity flight. Astronauts and jet pilots have been reporting on the effects of low or no gravity since the 1940s. But try to imagine what life is like without having any sensation of your body. It's impossible. Our identity of self, what it means to be an individual, is so intimately bound up with our physical bodies that it is impossible to imagine what the world would be like without proprioception.*

Ian's specific condition† is so rare that it affects fewer than a dozen people on the planet. But even within this exclusive group, Ian is exceptional because he innovated a way to compensate for his lack of proprioception by using vision. That discovery has made him famous within certain physiologic circles. He has become, in a way, the Rosetta stone of proprioception—the key that unlocked scientific and, indeed, human understanding of how this relatively unappreciated sense affects everything about us. And

---

*Indeed, some philosopher-scientists have questioned whether there would even be a sense of self without proprioception. Would you still be *you*?

†Ian Waterman was eventually diagnosed with acute sensory neuronopathy with near-total destruction of proprioceptive and sensory nerves due to an autoimmune reaction. Sensory neuronopathy of varying degrees can also occur as the result of a toxic level of vitamin B6, some cancers, and some doses of cisplatin chemotherapy.

very few people would know about Ian Waterman at all if not for Dr. Jonathan Cole.

The Indiana Jones of Proprioception

Dr. Cole is currently a consultant in clinical neurophysiology and psychology at Bournemouth University in the United Kingdom, but when he first met Ian more than thirty years ago, he was a "lowly research fellow."* Since then, Dr. Cole has been Ian's friend, biographer, research collaborator, and physician, in that order. In 1991 Cole published *Pride and a Daily Marathon*, a riveting description of Ian's condition and his physical and emotional adaptation, which has allowed Ian to find steady employment, get married, and generally have a life. Dr. Cole is an integral part of Ian's journey. He is the translator of the neurologic story Ian's condition is telling. So much of what we now know about proprioception is due to the bond between these men as they struggled to define exactly what was happening with Ian Waterman's body and mind. I began an e-mail correspondence with Dr. Cole with the hopes of understanding more about how their relationship began and how Ian's condition has contributed to our knowledge of how proprioception contributes to balance.

Despite much prompting, I could not get Dr. Cole to break his reserve. Perhaps that is due to having lived the narrative for several decades, or maybe it is simply a habit borne of a lifetime of describing the lives of others, but there are no emoticons in Dr. Cole's writing style. However, he was more patient and gracious than I had any right to expect. He is also quite humble, insisting that Ian deserves the attention and kudos. But for years, Dr. Cole was a neurologic archeologist, slowly and methodically helping to

---

*His words, not mine.

reveal more of the secrets of Ian's condition. Without Dr. Cole, Ian's remarkable story and achievements would be lost to science. According to Dr. Cole:

> When I realised how extraordinary Ian's condition and his response to it were, I had a mixture of feelings. Excitement at being in a position to investigate this according to how ingenious I could be and how enthusiastic for science Ian was, but also a sense of responsibility. . . . His story was so important I felt that I must not mess it up. He deserved the best. I just hoped I could do him and his story justice.

Dr. Cole first met Ian when a colleague called to say there was someone with a medical disorder he might be interested in. He reports that Ian was not keen on doctors but agreed to stay behind to meet this new physician. In Dr. Cole's words:

> When [Ian] explained his problem I was amazed; it was not something I had ever seen nor known could exist. His problem was spectacularly rare and yet so pure—absence of touch and proprioception, but with normal pain and temperature perception, and with the movement or motor nerves intact. I was already thinking of how to prove this before moving onto other experiments, if Ian was happy to come on board. Yet, immediately I could see an apparent mismatch between the severity of the clinical problem he described, and how he moved and walked. I would have anticipated he would have been wheelchair-bound with little controlled movement, and the usual irregular uncertain movements those without [nervous system] feedback have, so-called ataxia. He had none of these. Why?

The answer to that ultimate question is what makes Ian Waterman unique in the whole of sensory processing history.

Ian deliberately retaught himself to move by staring at a part of his body, creating an image in his mind of what moving it would look like, and then *willing* that movement to happen. The story of how this happened reminds me of my high school fascination with telekinesis—the ability to make objects move with the power of your mind.* I remember engaging in telekinesis training sessions, staring at a book, burning it with my thought-rays, trying to make it move, to nudge it just a little with the power of my mind. When that didn't work, I rationalized that maybe I had been too ambitious, that a book was too heavy. So I stared at a feather. It never worked. I imagine that Ian Waterman's first attempts might have felt like striving for telekinesis, trying to move an intractable body part by sheer concentration. But his personal training sessions were all he had. Still, nothing happened for the longest time. Hours. Days. Weeks. And then—he moved.

Ian's first success—sitting up from a reclined position—took several weeks to accomplish.† Learning to stand took a year. Over the succeeding years, Ian trained himself to approximate all normal movements in this ingenious way: by imagining or visualizing the movement he wanted while simultaneously looking at the corresponding body part. His brain finally transmitted those visualizations as commands to his limbs. It was a feat so amazing that it could be considered a form of neurologic telekinesis. So why was he able to succeed when my book-moving powers failed so dreadfully? Dr. Cole believes that two main factors were at play.‡ First, Ian was young at the time of his devastating illness. Most of the other people similarly afflicted were struck in their forties, fifties, or

---

*My interest, and that of several of my friends, was inspired by Stephen King's novel *Carrie*.
†He was so astonished by his success that he promptly fell over again.
‡Three, if you consider that real telekinesis doesn't exist, and therefore I never stood a chance of foiling the reality of physics.

older. Ian was nineteen. Teenage brains are still developing, building brain cells and creating networks of new connections. Though the sensory wiring from Ian's body to his brain was destroyed, his brain was capable of shoring up his imagination and visualization capabilities, which meant that this new way of initiating movements could be wired in. It may not have been possible if he had been middle-aged at the time of his illness.

The second factor in Ian's success was, as Dr. Cole puts it, "sheer persistence and bloody-mindedness." Any new physical skill takes time and practice to develop. But most of us can see the rewards of our efforts in incremental improvements, which spurs us to continue practicing. Ian had no such reward feedback. He simply soldiered on with no promise that his efforts would pay off. For him, the options were keep trying or become resolved to being wheelchair-bound or bedridden for the rest of his life.

Today, Ian can do many things a person with functioning nerves can do, despite the fact that he still can't feel or sense his body. He has learned to compensate for a lack of proprioception by using visual cues. He can look at his hand, imagine a movement, and make it move. But the movement is dependent on both the looking and the imagining. If the lights are turned off so he cannot see, or if he becomes distracted so his concentration fails, Ian's body falls in a heap, like a marionette whose strings are snipped. Ian even has to sleep with the lights on; in the dark he would be unable to turn from side to side or even to recognize that the object lying on his forehead is his own arm. He cannot fish change out of his pocket because his hand is out of sight. Ian also cannot do anything that requires speed, such as running or dancing. As Dr. Cole relates in his book, every moment of Ian's life is an intense marathon of concentration that is totally dependent on being able to see.

After their initial introduction, Dr. Cole became fascinated with another aspect of Ian's condition: What did it feel like to be Ian, to be missing proprioception? How did he manage in everyday life? Those were the crucial questions that bonded patient to physician. Apparently, in the twelve years since Ian's acute illness, no one had ever asked what life was like for him. No doctor, nurse, or physical therapist. Ian had been hungry for someone in the medical community to recognize him as a human being and not merely a case study. With that connection established, Ian agreed to what would become a lifetime of experiments, and to Dr. Cole's endless questions about his process for daily living.

As a result, Dr. Cole has published dozens of scientific articles and book chapters about facets of proprioception that are only known because of Ian Waterman's unique response to his disability. Two entire books have been devoted to Ian's progress.* On a grander scale, Ian's rehabilitation process could help guide a new way of thinking about recovery time for anyone with an amputation or spinal cord injury. Ian's initial recovery took two years, but he was still improving and experimenting with his abilities for years after. As Dr. Cole notes, in people with catastrophic injury, rehabilitation can take "months and years as people come to terms with their altered embodiment and find ways to become anew, to become all they are capable of." Through his persistence and patient practice, Ian has accomplished far more than anyone had imagined.

Pride and a Daily Marathon (MIT Press, 1995) and *Losing Touch* (Oxford University Press, 2016). These books are so much more than a view of a medical miracle. They also document a rare and touching friendship. Dr. Cole was the best man at Ian Waterman's wedding a few years ago, and he speaks fondly of their research travels together. I think it is safe to say that the relationship benefitted both men in ways that extend far beyond scientific inquiry and rehabilitation.

Balancing on a Leg That Isn't There

One of the lessons learned from Ian Waterman was exactly how difficult it is to keep balance without proprioceptive senses. As an example, consider this: When Ian stands, he has to lock his hip and knee joints so his body remains rigidly steady, and he plants his feet about shoulder-width apart, feet angled slightly outward, so they form a firm base. He cannot feel the ground, nor does he experience the physical pull of gravity to let him know that he is properly planted. Rather, the ability to stand requires continuous conscious thought while he simultaneously looks down to keep his lower body parts within his vision. While most of us employ an unconscious combination of vestibular, visual, and proprioceptive cues to maintain balance, Ian doesn't have that luxury. He has substituted continuous mental imagery and visual feedback for the missing proprioceptive signals. Simply balancing on his own two feet is a tedious miracle.

Lifting an object while standing is significantly more complex. Remember, Ian can't feel his body, and he can't feel anything else, either. Imagine standing with your eyes closed. If someone were then to put an object in your hand, you could easily tell the difference between an egg and a baseball, or a coffee mug, or a flowerpot. Ian cannot tell the difference unless his eyes are open—he can't sense the object's shape or weight. To lift the object, Ian needs to estimate its weight so he can make postural adjustments to account for the balance shift that the extra weight might cause. If the object is significantly heavier than he calculated, it can pull him over.*

---

*In addition to weight, an item's fragility is also an issue. Because Ian cannot feel objects, he also has to visually gauge when his grasp is appropriate to hold a particular item. Eggs used to be a particular challenge because if his grip was either too tight or too loose, the outcome was messy. However, he has found different strategies for complex tasks. Eggs, he learned, must be cradled instead of grasped.

Walking presents seemingly infinite opportunities for failure. Surface texture and pitch must be navigated, and carpeted floors are different from grass or dirt. Pebbled walkways can be difficult for Ian—the uneven surface and potential for slipping on a rolling stone are as treacherous as a slick and icy parking lot can be for the rest of us. Each step requires Ian's full concentration. He lives every moment at the edge of what is possible, like an elite athlete.

In order to keep his balance, Ian has developed a unique gait. He looks downward to maintain visual contact with his feet and legs and the ground. He stands with locked hips, knees, and ankles. Then he seems to simultaneously loosen control of his knee (allowing it to bend) and shoot the leg out from the hip. When his foot hits the floor, he locks his joints up again. As he shifts his weight to lift the other leg, the hip stiffens to accommodate the change. Stiff joints, loose joints, stiff joints. To an outside observer, it's an alien approximation of human walking. To Ian, it is mastery over a disorder that tried to tame his spirit. To Ian, it's freedom.*

---

*The BBC did a great documentary about Ian Waterman called *The Man Who Lost His Body*, available on YouTube and other sites online: www.dailymotion.com/video /x12647t_the-man-who-lost-his-body-bbc-documentary_tech.

5

Self-Orientation

The Gravity of Up

HAVE YOU EVER HAD A DREAM OF FLYING, soaring high up over the rest of the world? How did you know which way was up? In a dream, there is no gravity weighing you down, no change in the balance centers of your inner ear because you remain motionless on a pillow. Perhaps it was because you could see the ground below, so up was the direction opposite the ground. Fair enough. Imagine instead that you fly so high that you leave Earth's orbit, you leave the solar system, and you are flying among a vast expanse of space and stars. Now which way is up?

Balance begins with knowing which way is up. When you awake in your bed and open your eyes, you know that you are lying down and that the world hasn't suddenly shifted ninety degrees. It might seem self-evident, but it actually takes a combination of cues to allow you to come to those conclusions. First your brain gets cues from the vestibular system, especially the gravity-sensing tiny crystals of the inner ear, which are pulled downward toward the earth regardless of your body orientation. Second your eyes visually scan the environment for more clues—the fan is on the ceiling, so you know, logically, that up is where the fan is. And third your proprioceptive system tells your brain how your body parts are oriented in relation to each other and to the rest of the world.

Here's another example: Imagine that you had too much of a celebratory night and woke up dangling over the edge of a couch. With gravity pulling you (and your ear crystals) toward the floor, your vestibular system sends signals that you are upside down. When you open your eyes, the furniture would appear to be above your head, but—logically and from all past experience—you know that furniture rests only on the floor, so you must be hanging upside down. And your proprioceptive system tells you that your limbs are, unusually, above your head. Taken together, those clues allow you to recognize that you need to carefully turn yourself right side up again (and that you're in for a doozy of a hangover). Obvious, right? Right!

This sensory portion of balance is so obvious that it is virtually neglected in medical and psychological textbooks. And yet it is a fundamental part of how we experience and interact with the world. It is so basic that everyone takes it for granted. For example, if you're describing a party scene, you would never bother to mention the fact that all the people in the room were positioned with their heads vertical to their bodies and their feet on the carpet. And yet when we look at the world or at other individuals, the first thing we do is orient ourselves so that we line up with the up-down axis. We do this by what scientists call multisensory integration.

That is the research specialty of the brilliant Dr. Laurence Harris, professor of psychology, kinesiology, and health sciences at York University in Toronto, Canada, and founder and editor in chief of the journal *Multisensory Research*. His research examines what happens when the three orientation clues—gravitational, visual, and proprioceptive—are in conflict. In other words, what if what we *see* doesn't match the pull of gravity? What does that do to our sense of which way is up? His research is all about measuring the effects of sensory illusions.

My first impression upon meeting Dr. Harris is that he is chummy. From our first hello, I feel as though I have known him for years. He is that comfortable friend you invite everywhere because he gets along with everybody. He is rumpled cotton and mussed hair and worn leather loafers with no socks. He looks like he's always suppressing a smile, especially when he talks about his work with mind-blowing vestibular illusions. It's obvious he finds his research as cool as I do—like he lives to elicit the words "oh, wow." In total, his appearance and demeanor remind me of a slightly disheveled Puck or a light-haired Loki, perhaps, but without the dark mischief . . . although, now that I think about it, I wouldn't bet against Dr. Harris's being capable of pulling some shenanigans.

We linger for a moment in his graduate students' office. Desks, papers, cardboard boxes, books, and arcane electrical equipment litter much of the space. A large fish tank, at least a hundred gallons, hulks on the left wall. It houses three turtles, each the size of a luncheon plate, grown up from the quarter-sized, bright-green hatchlings that used to be a pet shop staple.* The turtles have been with Dr. Harris for about twenty years, and he shares the joys of feeding them and cleaning their murky tank with his PhD

---

*Baby turtles used to be sold in every pet shop and five-and-dime store (kind of an early Wal-Mart, with all manner of goods jammed into a space the size of an average school cafeteria) until it was discovered that handling turtles was a great way to contract salmonella poisoning. At that point, in 1975, the United States government outlawed the sales of turtles with shells smaller than four inches long. (The author is not admitting to remembering those long-gone days. Would you believe that she is a fan of reptile history and lore? Let's go with that.)

students.* Along the wall by the door is a couch reminiscent of other student sofas I have seen throughout North America—lumpy, saggy, covered with a quilt, and redolent of some familiar aromatic herb that had probably been smoked nearby. This is the entrance to the Multisensory Integration Laboratory. After a brief stop here, my fleet-footed guide launches into a tour down a rabbit warren of illusion experimentation.

I lose track of all the rooms that comprise the Harris lab. The Multisensory Integration Laboratory has at least six subrooms off what I would come to call the Turtle Room. Many hold wooden or electronic remnants of completed, postponed, or abandoned investigations, as well as gnawed foam rubber matting, raw wood sheets and two-by-fours, plastic clips, metal cages that could have been built with a large-scale erector set, and wires . . . wires everywhere. Wires that connect small boxes to the wall, wires that hang unconnected from old computers, wires from electrodes, wires attached to cameras, wires taped to the walls and floor, wires dangling from the ceiling. What strikes me first is how crude and rough the scientific devices look. Even the Barbie doll sitting on a wooden shelf is half bald. (She used to serve some useful purpose but is now just another messy remnant.) Much of the good stuff has been cannibalized to build the next great thing.

Dr. Harris explains that when the lab was started, funding was minimal, so everything was put together with "string and sealing wax." But as the research began to yield unique and valuable insights into sensory perception, the lab kept growing and grant money increased. The "string and sealing wax" devices are still there, but so is the state-of-the-art virtual reality theater. Still, some of the best sensory illusions are homemade.

---

*Strictly a voluntary chore, for which, I'm told, the students are paid.

The Reorientation Illusion

I walk into a room-sized box decorated to look like a room. A dreary, poorly decorated room. The instructions are to stand with my back against the right wall and examine, really notice the room. I try to remember the details because it feels like there might be a test. I first notice the mustard-gold wallpaper patterned with tiny trees and African animals—elephants, zebras, giraffes, leopards. Below the bright-white chair rail is a border of matching wallpaper, and below that is a kind of bamboo-texture pattern, also in mustard-gold. The floor is industrial brown carpeting. There's a tall window to my right. Against the left wall is a poster of mountains under a crystal blue sky with fluffy white clouds. There are electrical sockets and light switches. I stand there for several minutes and commit the room to memory, then leave the room.

Next I close my eyes while Dr. Harris leads me into a second room. I'm asked to lie on the floor, arms down at my side, with my heels in the seam where the wall meets the floor and my feet flat against that adjoining wall. I open my eyes to find that I am in a dreary, poorly decorated room with mustard-gold wallpaper patterned with tiny trees and African animals—elephants, zebras, giraffes, leopards. Below the bright-white chair rail is a border of matching wallpaper, and below that is a kind of bamboo-texture pattern. Exactly the same as the first room but with everything turned ninety degrees so that the industrial brown carpeting is under my feet (but on the wall), the poster on the left looks upright, and the tall window is still on my right. I know I'm lying down, and when I close my eyes I can sense gravity pulling downward, but with my eyes open—holy smokes!—it feels like I am standing upright. Eyes closed, lying down. Eyes open, standing.

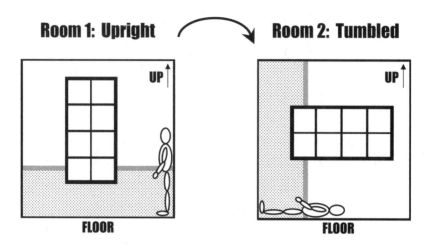

What happens in the Tumbled Room. In the first room,
you stand against the wall, with the room normally oriented.
In the second room, you lie on the floor, but the room's construction
and decorations are set at a ninety-degree angle,
giving the illusion that you're still standing on your feet.
Based on the work of Dr. Laurence Harris

I can't stop saying "wow!" It is mind-blowing. I feel as though I am not only standing up but also hovering! It's a sense of not quite weightlessness but . . . levitating. The feeling makes sense—gravity isn't pulling all my body weight down through my feet. Instead, gravity is at my back. But with my eyes open, I can't sense anything more than the "wall" against my back. Even when I remind myself that I am lying down, my brain is doing its own set of perceptual calculations and sends back the message, "Nope, you're standing up."

Rock, Paper, Scissors

My brain can be fooled like this because of how balance is wired in. In earlier chapters, I spoke about how the signals from the semicircular canals and balance organs in the inner ear transmit to the brain from the vestibular nerve, which feeds into the brain at an area called the primary vestibular cortex. Until relatively recently, scientists had assumed that this portion of the brain was *the* area responsible for balance. On the surface, that makes sense— the visual and auditory systems both have dedicated areas of the brain that are responsible for sense processing. For example, our eyes send nerve signals straight to the primary visual cortex at the back of the brain. There, nerve signals are decoded and translated in such a way that we *see*. That one portion of the brain is responsible—and necessary—for sight. A person can become blind if the visual cortex is injured, even if the eyes are healthy.

Balance is different. There isn't a one-to-one correspondence to a specific area of the brain. Several parts of the brain receive direct vestibular projections from the thalamus, which is sort of the sensor relay module for nerve impulses related to balance. Those areas of the brain are special because they don't respond *only* to vestibular information—they also respond to visual and proprioceptive information.*

It turns out that these multiple sensory inputs affect and define our perception of up. From the body, we get proprioceptive information (knowing where your head and body are located in space). We also have the long-standing, learned-from-infancy assumption that our head is up and our feet are down. That immediately gives you a general reference point for your orientation. From gravity, we can feel the pull toward the earth, defining down. And from vision, we gather external cues such as the direction of the sky or

---

*Interestingly, while the visual cortex is wholly responsible for vision, it also receives input from other sources. For example, the cells in the visual cortex also respond very strongly to proprioceptive information, the position of the eyes, and that sort of thing. Some even have auditory input. The whole idea of dividing the brain up into specialized areas that do just one thing—just vision, just hearing, just balance—is a bit of a misconception that may have evolved from the need of textbook writers to be able to divide information into neat, self-contained chapters.

a ceiling, which we know almost always indicates up. In the case of the Tumbled Room illusion, I had spent a decent amount of time in room 1 learning the exact visual references for the direction of "up" by closely examining the decor. That's why the initial experience of the identical right-side-up room is important—it provides a visual and cognitive anchor. Without it, the illusion wouldn't work consistently, and it certainly wouldn't be as strong. Combine those three factors—body, gravity, and vision—and you've got a pretty infallible and universal sense of orientation.

Dr. Harris and his colleagues have been able to determine the relative contribution of each of those cues: 50 percent comes from body cues, 25 percent from gravity cues, and 25 percent from vision cues. Rock, paper, scissors. Body, gravity, vision. Which one wins? In each situation, our brains combine all the information, do a quick calculation, and make a determination of which way is up. The reorientation illusion messes with the percentages, virtually forcing the brain to make an incorrect calculation. I saw a room with the "ceiling" over my head. Visual (25 percent) and body (50 percent) cues both pointed to one direction of up, while gravity (25 percent) was the only outlier. So, in the Tumbled Room, 75 percent of inputs told me that I was standing up, so that's what I experienced. When I closed my eyes, I no longer had mustardy wallpaper to confuse my brain, so gravity was able to play its natural role and remind me that I was lying down. As long as I had my eyes open, I felt as though I were levitating upright in the Tumbled Room. Very cool illusion.

3-D Vection Illusion

In another part of the York University campus, down a graffiti-marred stairwell, is the basement home of the Tumbling Room—an eight-foot cube held several feet above the floor by thick metal stanchions. I'm not mechanically inclined, but I take note of multiple crossbeams, gears, pulleys, counterweights, cords, wires, brackets, clamps, and sundry other builder-type parts I can't identify. It's an overwhelming number of interconnected components.

The suspended box is another room, and the permanent home of Hans, a tanned and classically handsome mannequin that looks to be of the 1970s vintage with long, beach-blond hair and cheekbones so sharp they could gouge an eye. He is seated, posed leaning forward on a table set with bowls, cup, saucer, utensils, and a basket of bread. There is, again, animal-print wallpaper, but this time it is a barnyard theme. Someone spent a lot of time attending to homey touches throughout the room—scarf casually thrown over the back of a chair; multiple pictures hanging on the walls; and shelves decorated with hanging teacups, books, and a basket of silk pansies.

These details are key to the illusion.

Hans in the Tumbling Room. Note all the details.
Carol Svec

On the opposite side of the room is my seat—a large wood, metal, and foam cradle. After I duck and settle into the frame, Dr. Harris immobilizes my head between duct-taped cushions and secures the rest of me with a five-point pilot harness from a fighter jet. Snug as a bug in a . . . wait a second. It is at this point that I ask exactly what will happen and why I need to be strapped in so tightly. My roguish guide tells me that the illusion will be more potent if I don't know anything about it beforehand. As Dr. Harris leaves the room, I give him a big smile and two thumbs up. Then it's just me and Hans, who, I notice, is not strapped into a giant chair, *and* who appears to have gone through some sort of personal trauma, what with his torn shirtsleeve, disheveled hair, and haunted expression. I won't lie to you—I'm having second thoughts.

After a few long seconds, I begin to tumble. Rather, the chair I am strapped into begins to rotate to my right. I turn sideways, upside down, sideways (to the other side), then upright again, over and over. It is a trip back to childhood, as I feel the chair doing cartwheels. After a couple of exhilarating rotations, I notice that my hair isn't moving. And there are no signs of a change in the direction of gravity—if I'm really upside down, I should feel a pressure from the straps across my shoulders, but I don't. It occurs to me that maybe the room is moving around me, but there is no visual evidence of this. Every hair on Hans's head stays perfectly in place, the pictures on the walls don't sway, the scarf and teacups don't budge, and the dishes and utensils remain firmly on the table. It feels like gravity is broken.

That's when I begin laughing. This illusion is delightful! I seriously can't objectively tell whether my chair or the room is moving, but it feels like it must be me. I'm tumbling. No amount of concentration can convince my brain that the room is moving; I simply can't sense it. I'm cartwheeling! I keep laughing until all movement stops and Dr. Harris opens the door again.

"That's a good one, isn't it?" he says.

I agree. And I have a fleeting desire to move in with Hans and have this experience every day.*

Going with the Optic Flow

According to Dr. Harris, the Tumbling Room has been used by NASA to train astronauts to deal with disorientation cues while in outer space. The illusion is a combination of the Reorientation Illusion (in that my experience of gravity was altered while the room was rotating) and visual vection. With visual vection, we get the illusion of self-motion when a large or obvious portion of the surrounding environment moves unexpectedly. But instead of recognizing that the *environment* is moving, we get the sensation that *we* are moving. Scientists are still questioning why this happens and what significance it has for our everyday lives. The most popular view is that this illusion is an artifact of our more primitive selves, before we got tricky with psychological and physics experiments.

As we move through our environment, either as prehistoric hunters on foot or bargain-hunting shoppers on the way to the mall in a car, we are able to navigate to our chosen location based on the complex information we receive about the three-dimensional world around us while we are moving. This is called optic flow.

You've experienced optic flow without even realizing what it is. When you're in a car driving on a straight stretch of road, the objects far in the distance appear small and almost motionless.

---

*That probably wouldn't work out so well. I spotted Hans's last roommate, Marta, a green-eyed, rail-thin female mannequin, hiding out across campus behind a jumble of excess furniture. Her brown wig was terribly askew. Oh, Hans, what did you do?

However, as objects get closer and closer to you, their images get larger. (This is another illusion, albeit one we learn to ignore in childhood, a concept known as size constancy. We know that the objects don't actually get bigger—only that *bigger* means they are closer to us.) And as you drive past an object, it appears to approach faster and faster until it whips past your side window at a very high speed. This pattern of how we perceive objects as we move through the environment is called optic flow.* It helps us understand the structure of our environment and allows us to maneuver safely without running into a tree. It has been suggested that humans and animals have brain cells dedicated to the computation and translation of optic flow information. That's why we don't get freaked out when objects appear to grow before our eyes and streak by at impossible speeds. It's also why an alert driver can (sometimes) steer around a squirrel that runs into a road, or avoid a chair that falls off the back of a pickup truck in front of her.†

Those dedicated brain cells were developed way before cars, and yet we are able to translate the information from one form of forward motion (walking) to another (driving). In both cases, we are moving through the environment in a manner that we control.

What nature never saw coming, though, was the Tumbling Room. Two things (at least) are happening there to create the

---

*Another example of optic flow is a moving star field, in which dots of light (stars, presumably) appear to move toward and around you. You can see an example of a "warp speed" star field pattern here: www.youtube.com/watch?v=iJLbAntU118.

†Yes, I have avoided the flying highway chair. And even though I know it is dangerous, I automatically try to steer clear of squirrels. I can't help it. It's like my primitive brain recognizes the object hopping innocently into my path and instinctively reacts to avoid it. One time a car was coming toward me in the opposite lane when a squirrel ran out from the side. My brain was smart enough to choose not to swerve into the large metal vehicle, but I did tense up for the expected *thump-thump* of the squirrel under my wheels. No *thump-thump*. I looked in my rearview mirror to see the squirrel bouncing back into a tree. I'm sure that our individual optic flow calculations saved both of us.

illusion. First, visual cues are telling me which way is up, *and* that up is moving. Second, optic flow registers that the environment is rotating in a circular path. But our brains are wired so that optic flow registers how *we* are moving through the environment. With no reliable visual cues to tell me otherwise, my brain extrapolates that I must be cartwheeling around this stationary room.

"Up" in Space, or How to Make Sure You Don't Open the Wrong Hatch

As my experiences with the Tumbled and Tumbling Room illusions illustrate, figuring out where up is can be manipulated pretty easily here on Earth. In space, astronauts face an entirely different set of challenges. When I was in Dr. Harris's illusion rooms, if I closed my eyes, the illusions were broken. Without the compelling visual trickery, my body could sense the pull of gravity and I could once again be able to point to true up. In space, you are weightless. Close your eyes in weightlessness and there is no gravitational pull, no directionality.

In space, there is no up.

At first think, that may not seem like such a big deal. But, in fact, not having a solid and reliable reference point for directionality can mean the difference between mission success and critical error.

Though astronauts are weightless and without a reference direction, some features of a spacecraft need such orientation clues. On a very basic level, some tasks require that a switch be flipped *up*. Therefore, some kind of special directional information needs to be present near the switch to avoid disaster. Researchers have also found that it is more difficult to navigate around larger areas (such as the International Space Station), to operate equipment, to read signs, or to recognize objects and even colleagues' faces when they are not in their usual orientation relative to the observer. Have you ever watched someone talk while viewing them upside down so that her mouth is above her eyes? It's amusing, but it also demonstrates how something as common as a face can look alien when viewed from a skewed perspective.

Thanks to Dr. Harris, we know that on Earth our sense of up is determined by a combination of 50 percent body cues, 25 percent vision, and 25 percent gravity. What happens when there is no gravity? Which cues become most important? Knowing the answers to those questions will help NASA and other space organizations to design equipment that will minimize the chance for errors in space. So Dr. Harris and several of his colleagues went into almost-space—they took a ride on a modified Falcon 20 and experienced weightlessness in twenty-two-second bursts.

The average person can experience microgravity (near weightlessness) via the magic of parabolic flights, one of the techniques used to train astronauts and pilots. What happens is this: The plane takes off and reaches a steady cruising height. Then it climbs at about a forty-five-degree angle, and the pilot accelerates (throttles up). During this climb, gravity is increased so that your body feels about 80 percent heavier than usual. That means a 200-pound person will suddenly feel 360 pounds. Just lifting your head will feel like a monumental task.* This is called hypergravity. Then, the pilot throttles back and the plane levels out. At this point you feel weightless, capable of floating around the cabin the way you see astronauts "hanging around" in images sent from space. This joyous freedom lasts only about twenty-two seconds. Then the plane

---

*There have been rumors that some dudes perform push-ups during this portion of the trip. All I can say about that is that nobody likes a show-off! Also, can you be my personal trainer?

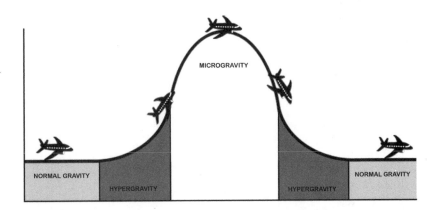

How microgravity happens in a parabolic flight.
Carol Svec

dives back down at about a forty-five-degree angle and cruises
steady for a bit, before repeating the process, sometimes up to
forty times. If that sounds like a stomach-heaving experience, it
is. According to NASA, somewhere between 60 and 80 percent of
parabolic passengers get sick their first flight.*

By the way, you can get a similar experience by riding a large
roller coaster with a parabolic path or steep vertical drop, such as
the 310-foot-tall Millennium Force or 420-foot Top Thrill Dragster,
both at Cedar Point Amusement Park in Sandusky, Ohio, or the
456-foot Kingda Ka at Six Flags Great Adventure in Jackson, New
Jersey. When you reach the topmost portion and begin descent,
you'll be temporarily weightless, and when the coaster is climbing

---

*Dr. Jonathan Cole and Ian Waterman, the man described in the previous chapter who
lost all proprioception below his neck, went on one of these flights. Dr. Cole reports that
Ian did not experience any extra heaviness during the periods of hypergravity, nor did
he feel any differently in the weightlessness portion. Because of his condition, he always
feels nearly weightless, wherever he goes, whatever he does. The flight did not impress
him, except that he was one of those who got quite motion sick.

Millennium Force roller coaster at Cedar Point in Sandusky, Ohio.
Cedar Point

upward or going around a bend, you'll be pressed into your seat with a form of hypergravity.

So that was what Dr. Harris had to work with to find out how people define up when there is no gravity—twenty-two seconds of weightlessness, four times per flight, four flights in a day. That's less than six minutes of actual experimentation time (but a full day's worth of motion sickness). Still, the results were intriguing. Remember: on Earth, we use body, vision, and gravity cues when deciding which way is up. The results of Dr. Harris's research showed that when gravity was taken away during the parabolic flight, test subjects gave vision cues less weight and used primarily body cues. That is, weightless subjects defined up as whichever way the top of their head was. This becomes important when you consider how disorienting space flight is and how not knowing which way is up could mean pressing the wrong button, flipping a switch the wrong way, or any of a million other directional mistakes.

Tripping on the Moon

Astronaut Harrison H. "Jack" Schmitt is one of only a dozen men to have had the privilege and honor of walking on the moon. Even more impressive, Schmitt was a geologist and was chosen for astronaut training because of his scientific expertise; he was the first and only moonwalker who didn't come with a pilot pedigree—a remarkable achievement. And yet Schmitt's legacy in this social media age is his difficulty remaining upright when walking on the moon. At one point, after Schmitt toppled yet again while collecting rock samples from the moon surface, the Apollo 17 Mission Control asked Commander Eugene Cernan to "go over and help Twinkletoes, please." The nickname stuck.*

The question of why so many astronauts—no, not just Schmitt[†]—had trouble remaining upright on the moon nagged at Dr. Harris. Perhaps it had to do with being unable to correctly discern up. Based on the results of the previous experiment, Dr. Harris knew that the personal cues that give us a sense of up change dramatically once gravity is taken away. But exactly how much gravity is necessary to once again exert an effect? How much gravity is enough to help us define up?

As he always does when a question nags, Dr. Harris conducted another study. This one made Dr. Harris a rock star researcher as the media reports of his results went viral.[‡] To

---

*YouTube video that includes Mission Control calling Jack Schmitt "twinkletoes" is available here: www.youtube.com/watch?v=ZP7AVBdJYOg.

†You can watch a compilation of multiple astronauts falling on the moon here: www.youtube.com/watch?v=LEdYf4SGhuI.

‡Well, as viral as science news gets.

understand this study, imagine just lying on your back with a video screen hovering above your face. Just as with my experience in the Tumbled Room (the Reorientation Illusion), gravity will be at your back, pulling equally from head to toe. Therefore gravity won't have any effect on your perception of an image displayed on the screen. (As Dr. Harris puts it, it is the world's cheapest zero-G simulator.)

In this study, each test subject was asked to lie down on a centrifuge platform, with feet facing outward and a screen overhead. Against a tilted background image (to enhance a shift in perceptual upright), the letter *p* was displayed. Now, interesting fact about the letter *p* is that when viewed upside down, it is the letter *d*. Not an earth-shattering revelation but critical to this experiment—as the letter *p* varied in orientation, subjects would note whether they saw the *p* or the *d*. This sounds simple, but what if you are presented with this: ◌? You have to make a choice—is it a *p* or a *d*? When lying on an unmoving platform, the subjects reacted in ways that were similar to what you might find in a weightless environment. That is, they relied primarily on visual clues to distinguish *p* from *d*. (For example, if the *p* was presented exactly sideways, subjects used the background image to give them clues about the letter's orientation.) Dr. Harris then began spinning the subjects on the centrifuge and repeated the experiment at various speeds. Centrifuge spinning creates an acceleration effect, which adds back the sensation of gravity.* The faster the spin of the centrifuge, the more pull would be felt at the feet, simulating more gravity. During each level of increased speed, the subjects performed a test judging their perception of up.

---

*This is like when you're in a car and step on the gas. The fast acceleration makes your head feel heavier for a few moments.

The results showed that people only begin to use gravity to correctly define up when it reaches a level of about 15 percent of Earth's gravity. The moon has 17 percent of Earth's gravity. That means that the men who walked—and tripped—on the moon experienced a level of gravity just barely over the minimum. No wonder the astronauts—especially geologist Schmitt, who had less low-gravity training than other astronauts—had such a difficult time remaining upright. Among many other encumbrances, including the bulky spacesuit and equipment packs, the men could hardly tell which way was up. And without that grounding, balance is, at best, precarious.

6

Life-Changer

Persistent Postural-Perceptual Dizziness

MOST PEOPLE DON'T REALIZE how much vision contributes to balance. We understand that the vestibular system of the inner ear is the star of our balance system, and we have an intuitive sense of the importance of the body sensations of proprioception—that is, we know how our bodies feel when we are falling and we know where our feet are. We don't really think about vision. And yet, research has shown that vision can be more than just a second banana when it comes to balance—it takes up the slack when other systems fail. To anthropomorphize, vision is sort of the Eve Harrington of vestibular senses. For those of you who don't obsess over classic movies like I do, this is a reference to the 1950 movie *All About Eve*. In it, Eve Harrington is an avid fan of aging Broadway star Margo Channing (played by Bette Davis). Slowly, Eve begins taking over Margo's career and personal life. Eve orchestrates a plan: She becomes Margo's understudy, then arranges to have Margo miss a performance. Eve supplants Margo as Broadway's darling, finally taking the spotlight for herself.

Not that vision causes other vestibular senses to fail. But if any other portion of our sense of balance happens to become weakened or unreliable, vision can step up to claim the spotlight. In earlier chapters we saw that when there is no gravity, vision takes up the slack to become the dominant balance sense, and when there is no proprioception, as demonstrated by the remarkable life of Ian Waterman, vision becomes not only primary but crucial.

This built-in, fail-safe system protects our sense of balance, even under partially debilitated conditions. But what happens if vision is inaccurate? What if our eyes don't correctly register movement in the environment or translate the optic flow* that helps us navigate through the world? One possible result is a balance problem called persistent postural-perceptual dizziness (PPPD).†

People with PPPD may have all the usual symptoms of a vestibular disorder—dizziness, disorientation, nausea, and vertigo‡—but the symptoms are triggered by repetitive, moving, and/or complex visual patterns. Think, for example, about walking through the aisles of a grocery store where the shelves are piled high with multiple multicolored products. Imagine walking past boxes and bags and packages, aisle after aisle after aisle. That kind of slow-motion flow of repetitive visual stimulation makes some people sick.§ But moving through shelves of cereal boxes and laundry detergent isn't the only trigger for PPPD. Symptoms can also be

---

*Optic flow, described in more detail in chapter 5, is the term used to describe how images stream past us as we move through the environment.

†Until recently, the disorder was called visual vertigo.

‡Interestingly, and for reasons we don't understand, people with PPPD usually don't vomit, despite feeling nausea.

§The grocery store example is so potent that PPPD has also been called supermarket syndrome and sometimes SEE sick syndrome (SSS).

triggered while riding in a car with the sight of scenery zipping by the side windows, or from something as simple as watching the rotation of a large ceiling fan or walking on a crowded street (especially if you're bucking the flow of pedestrian traffic so that people are coming toward and streaming around you). If you're one of the unlucky ultrasensitive people with PPPD, such as my new friend Helen, symptoms can be triggered by nearly everything in the environment.

"It's a Lonely Existence"

Helen has been called crazy more times than she can remember, but not always with that exact word. Sometimes a person might suggest that Helen's illness is "all in her head" or that she's making it up or that maybe seeing a psychologist would help or that her symptoms can't be as bad as she claims they are. But Helen is not crazy, and she's not fabricating her illness. Helen has a robust case of PPPD.

Helen's dizziness and vertigo symptoms are off the charts, including nausea so severe that one of her health care providers recalls actually seeing Helen's skin turn green during a treatment session. For Helen, incapacitating illness is set off by any repetitive or complex visual motion. Stirring a pot of soup is impossible—the circular visual pattern makes her sick. Ditto for making meatballs. She had been an accomplished knitter until the repetitive motion of the needles became a trigger for her symptoms. (She has recently started knitting again, but only simple scarves or blankets, things she can create with her eyes closed.) Helen can't sweep the floor because the back-and-forth motion of a broom triggers dizziness; instead she keeps her house clean by dragging a Swiffer mop behind her where she can't see it. Reading is difficult because it requires back-and-forth eye movements, but using a computer is worse because of the brightness, random images, pop-up ads, and the need to scroll. She can watch TV in small doses as long as there isn't a lot of visual flux—daytime talk shows or how-to programs are fine, but action shows are out. James Bond thrillers are torture. Helen can't even take refuge in nature—her world starts spinning from vertigo simply by watching the falling snow or water running in a river. Helen can't negotiate stairs without dizziness and the fear of taking a tumble. She cannot drive. When she must go to a grocery store or mall, she wears sunglasses to reduce light and make the complex environments dimmer, and she carries a cane not only for support if a dizzy spell comes on but also as a signal for others to give her a little extra room.

PPPD has constricted Helen's life to a sliver of what the rest of us take for granted. She's not a complainer—during our conversation she kept apologizing for sounding so negative—but she summed up the effect of her disorder by stating, "It's a lonely existence."

Confusion Wrapped in a Question Mark

To learn more about PPPD, I spoke by phone with clinician and prolific researcher Dr. Susan Whitney, professor of physical therapy at the University of Pittsburgh School of Health and Rehabilitation Sciences. Her no-nonsense demeanor belies a deep compassion for the patients she treats.

According to Dr. Whitney, PPPD is a problem in search of answers. We don't know what causes the visual system to develop this kind of hypersensitivity or who is at risk of developing it. We don't have a good, definitive diagnostic test—a person might be said to have PPPD when her symptoms match the syndrome profile and only after all other vestibular disorders have been ruled out. For kids who are affected, going to school is a daily nightmare, since their vertigo can be triggered by something as simple as the visual overload of walking in a crowded school hallway between classes—think about the rush of other students and the seemingly unending rows of lockers. Too often, these kids become socially isolated. They can't enjoy the things that most kids do together—riding bikes, shopping at the mall, going to a fair or carnival, or just talking with friends between classes. It's tough to make friends when you may not even be able to watch TV together. According to Dr. Whitney, seeing the desperation of a patient living this kind of "lonely existence" spurs her on to help answer the riddle of the disorder.

There are three general theories about what causes PPPD.

The first theory suggests that signals to the brain from the visual world may become unreliable if there is something wrong with a person's vision—if the eyes are uneven, vision is tilted, or there is unusual nystagmus (automatic, jerking eye movements) or other eye condition. In those cases, complex visual scenes or repetitive movements can trigger a kind of motion sickness due to wonky vision creating left-eye/right-eye sensory conflict.

The second theory suggests that people with PPPD have greater visual dependence. Basically, that means that vision takes up more of the balance spotlight than the other vestibular senses. Those individuals rely more heavily on the sense of sight to judge which way is up, and to move within an ever-changing visual world. Typically, we weight vision as 25 percent of our balance-related information. For some reason, people with PPPD may have developed a personal system that depends on sight more than usual. Perhaps they had an inner-ear disorder when they were young and learned to compensate for vestibular system loss by having vision become more dominant. This strategy may have served them well at one time, but with all the stimulus of the visual world—movement, colors, repetition—the overabundance of visual signals in a visually oriented brain may not be decoded properly, leading to dizziness.

The third theory suggests the brains of people with PPPD may have difficulty distinguishing and processing central versus peripheral vision cues. To illustrate the problem, think about what happens when you're driving down the road. An attentive and relaxed driver scans the entire environment: road markings, traffic patterns, overhead signals, pedestrians on the sidewalk. Your eyes and brain coordinate to take in all that information in a way that allows you to make split-second decisions as you navigate a two-ton vehicle going forty-five miles per hour. You use your central vision (the bits right in front of you) and your peripheral, or ambient, vision (the stuff on the side you see out of the outside corners of your eyes) equally and in concert. But now imagine that the driver in front of you is swerving across lanes, and you decide to memorize the license plate. As you zero in on the bumper of that potentially drunk driver, you switch from a balanced use of your full range of vision to a focused use of central vision. That kind of central focus is great for allowing you to facilitate a DWI arrest, but it comes at a cost—by focusing on your central vision, the stuff on the sides becomes lost. That means that at the very moment you write down the final digit of the license plate, you are less likely to see the pedestrian stepping off the curb into your car's line of travel. That's OK—not for the pedestrian, of course, but that's normal visual function. Your eyes function seamlessly, taking in the whole of the environment but focusing more narrowly when necessary.

For some people, that seamlessly working visual system can develop more seams than a *Project Runway* audition wardrobe.

Vision doesn't just happen in the eyes—they are just tools for gathering information. The real magic happens in the brain, where all the bits of visual data are gathered and interpreted. Think of the eyes as cameras—fun equipment but useless without a memory card and software to transform the captured data into recognizable images. Optometrists believe that people with PPPD have difficulty processing visual signals in the brain. Going back to my driving-down-the-street example: Imagine trying to focus on license plate number (in your central vision), but it's hard to focus; you also see the leaf floating by your left window, the lemonade stand set up on the sidewalk to your right, the streetlights overhead, and even the movement of other cars reflected in your mirrors. Even though you want to focus on one thing, the process is broken and your brain gives equal weight to everything in your central and peripheral vision. Why? Again, no one really knows. Maybe it's due to a developmental problem that started in infancy, or maybe a concussion or virus affected the brain's processing centers. It's also possible that a pair of poorly constructed eyeglasses could create skewed or double vision. Regardless of the initiating cause, there seems to be a disruption in the way the brain decodes visual signals. That disruption can create nausea and dizziness.

For Helen, nearly all movement in the environment is out of sync, so it all makes her sick.

Looking back, Helen suspects she must have had PPPD all her life. When you're a kid, you believe your life, your sense of self, is normal because your personal experience is all you know. Helen's life experiences were different from most. She remembers not being able to stop in the school hallways to talk with her friends because of the risk of dizziness, and she could never learn to ride a bike—her sense of balance was always off. Helen always felt a little dizzy, which made common activities a challenge. For example, she had to force herself to endure going on an escalator or jumping off a diving board because she wanted to fit in with her friends. If they did it, she did it. She thought that everyone felt dizzy on an escalator, that imbalance was just part of life. Helen also remembers trying on her wedding dress, stepping up on the low platform so the seamstress could make alteration to the hem, and becoming uncomfortably dizzy. The disorder didn't become debilitating until approximately ten years ago. It's difficult for her to remember the details of how it all got off track because her life since has become enmeshed with the illness—it's all a blur of dizziness and vertigo. She doesn't really care *why* it happened. Her main question is how to fix it.

Watch the Shiny Disco Ball

Sadly, just as we don't know exactly what causes PPPD, scientists don't know how to cure it 100 percent of the time. Dr. Whitney's personal quest is to discover the who-what-whys of the disorder, not merely for the sake of advancing the science, but also so she can directly apply the information to people like Helen, who can't seem to find relief.

Dr. Whitney's specialty is physical therapy. Medicine has little to offer people with PPPD—perhaps a prescription to reduce fluid retention (for the fluid buildup of Ménière's disease, if that is a concern) or a benzodiazepine to help fight the panic that often accompanies attacks of PPPD. Hope can be found in physical therapy with a psychological concept called habituation.

The idea behind habituation is that, over time, we are capable of getting used to just about anything given frequent repetition of a trigger stimulus. Start with small doses of a mild form of the stimulus, and slowly increase the intensity until we no longer feel the offending symptoms. It's like the old story about the best way to boil a frog: If you put a frog directly into boiling water, it will immediately jump out; but if you put a frog in cold water and gradually increase the heat, the frog won't notice the temperature change and will stay put and be cooked in the pot.* This process works with helping to treat seasickness—experts say the only way to truly overcome seasickness is through habituation. Start by sitting in a docked boat just until you begin to feel ill, then get out. Come back the next day and repeat the process—sit for as long as possible without crossing the line into feeling full-blown nausea. Repeat until you can sit in a docked boat for an hour. Then, take the boat out and stay onboard until you begin to feel the symptoms coming on, then disembark. By experiencing small, controlled doses of the sickening experience, your body learns to tolerate being on a boat. The military uses habituation techniques to help troops tolerate air- or seasickness, and NASA uses it to help astronauts tolerate space sickness that comes from weightlessness.

Dr. Whitney and other vestibular physical therapists use habituation to help people with PPPD overcome their responses to the individual stimulus that sets them off. For example, Dr. Whitney treated a police officer whose PPPD was triggered by the flashing lights of his cruiser when he responded to calls at night—inconvenient, to say the least. It got so bad that the officer worried about eventually being unable to do his job. Dr. Whitney's solution required three main props: sunglasses, a dimmer switch, and a lighted disco ball. Yes, a disco ball. No longer a bygone symbol

---

*This is an old and well-known story, but it still gives me the willies to write about the slow death of what I imagine to be a cute, big-eyed frog. So it was a relief to learn that this story is nothing but a myth. According to Australian science guy Dr. Karl Kruszelnicki, frogs have a maximum level of heat they are willing (and able) to withstand. Once that temperature is reached, they will actively seek escape and will jump out of the water.

of the 1970s, disco balls have come back in style again, at least for treating PPPD. These tools aren't the same giant mirrored balls that used to hang from the ceiling of just about every decadent dance club. There are smaller versions that sit on a table and emit light in a rotating pattern. It's a good approximation of, among other stimuli, police cruiser lights.

While treating the police officer, Dr. Whitney's first step was to determine his sensitivity level. If he had been measured on a scale of 1 to 10, his sensitivity level would have been a 15. The officer became sick with even a minimal amount of light movement. At first, he could tolerate only the tiniest amount of light and only while wearing sunglasses to mute it even more. Slowly, over a period of months, Dr. Whitney worked with the officer to extend the amount of time he could tolerate the moving lights, increasing his exposure in each session, taking him just to the point of feeling sick before backing off.

This process requires patience. In order to extinguish a trigger, you must creep up on it very, very slowly. People who are triggered by the high shelves and visual stimuli of a grocery store can't just force themselves to go shopping every day and expect to be cured. That kind of immersion doesn't work—it's too much all at once. Instead, patients need to go through the stepwise process of habituation, gradually increasing tolerance of smaller stimuli before attempting full-on triggers. Why is that slow process important? Once again, as Dr. Whitney says, no one knows. Whatever is happening in the brain is still a mystery.

This type of treatment requires a tremendous commitment from patients—they know that every session will end with nausea, and they have to do it for months. As you might imagine, not everyone can go through with it. About half of patients drop out without completing the process. Physical therapists often recommend that patients watch specific YouTube videos that include strong trigger scenes for extra habituation sessions at home, in a comfy chair and at a time of their choosing.* Even then, patients don't tend to keep up with treatment. A person has to be highly motivated to choose to make themselves sick day after day, for months and months. The police officer had the necessary stamina, and he was a success story. Now he can chase bad guys all night long with the cruiser lights flashing, with no vertigo.

Experts believe that habituation might work by rewiring portions of the brain. Every new skill we learn builds new brain connections. Remember back to when you learned how to drive. In the beginning, you need to concentrate on every action, every stomp on the brake, every turn of the steering wheel. But with enough practice, driving becomes second nature because your brain created new communication pathways. Something similar might be happening with habituation. With enough repetition, we create new neural pathways. In other words, we can rewire the brain.

But treatment by habituation isn't universally successful. Some patients diligently go through the program and yet continue to experience vertigo. Helen's treatment wasn't successful. Dr. Whitney opened her whole bag of physical therapy tricks, but Helen remains incapacitated by her PPPD.† Why was the police officer's treatment successful while Helen's wasn't? Again, no one knows. PPPD is one giant scientific question mark.

---

*Some examples of videos chosen by Dr. Whitney's team for their triggering ability:
  Easy: Driving in Manhattan (5:10) www.youtube.com/watch?v=Lv0OgDmwC3Q.
  Medium: Walking in shopping mall (4:52) www.youtube.com/watch?v=L1q_6vKy2Qk.
  Difficult: Optokinetic Stimulation (10:13) www.youtube.com/watch?v=kAPtu1WTHYc.
  Hard: London Underground (11:19) www.youtube.com/watch?v=c1C3-0unDG0.
†During the course of the therapy, Helen learned that wearing sunglasses helps to make some visual stimuli more tolerable. She wears them anytime she has to leave the house, including inside a mall or grocery store.

What's Going on in There?!?

Dr. Whitney's latest research as a member of the team at the University of Pittsburgh Medical Center Eye & Ear Institute with Drs. Joseph Furman, Theodore Huppert, and Patrick Sparto takes her inside the brain with functional near-infrared spectroscopy (fNIRS). This has a little complexity, so bear with me.

All brain activity requires oxygen, which is delivered in blood via a network of four hundred miles of tiny capillaries. Everything you do engages a different part of the brain, and extra brain work requires extra blood flow. So when you plan what to have for lunch, try to solve an algebra problem, remember the name of your best friend from high school, walk to the refrigerator, listen to music, or anything else that requires thinking, moving, or communicating, the blood flow patterns in your brain change. With fNIRS scientists "see" what's going on in the brain by measuring changes in blood flow in response to specific stimuli or actions. For example, when you're searching your brain data bank for the name of a long-lost friend, fNIRS shows scientists what portion of the brain is involved in this kind of chore.

Because our brains are all wired pretty much the same, scientists can map the regions of the brain responsible for specific tasks. They can also examine two different groups of people—let's say, those with PPPD compared to those without—and see if their brains light up with activity in the same areas or if there are variations. If a difference is found, that's the big ah-ha moment. Once scientists understand what part of the brain is responsible for PPPD, they can start looking for better, more universally effective treatments.

That's what Dr. Whitney and her team are doing now. And it's a big, impressive team. Along with Dr. Whitney and medical expert Dr. Joseph Furman, there are physicists, computer experts, engineers, and even a navy captain PhD candidate. You'd probably have to go to NASA to find more concentrated brain power working on a single research question.

During studies, research subjects wear a cap covered in optical fibers, which look like regular wires or electrodes, at least to an untrained eye like mine. Half the fibers transmit near-infrared light, which penetrates the skull into the brain to a depth of about one inch. Near-infrared light is absorbed by high-oxygen hemoglobin, so more of the light will be absorbed in highly active parts of the brain. The other half of the fibers receive information about the amount of that light left after absorption by hemoglobin. By comparing the amount of near-infrared light transmitted and the amount left after blood absorption, scientists can tell which parts of the brain are most active.

In one of Dr. Whitney's current experiments, subjects wear the fNIRS cap while standing in front of a 180-degree bank of virtual reality projection screens that display an optic flow star field.* This is a common trigger for PPPD, and with the large screens, the effect is so powerful that even people without PPPD sway a little as they watch it. As the giant star field plays, the fNIRS cap feeds brain information to a computer. At the same time, subjects' balance is measured via sensors in the platform they stand on, and they report their subjective feelings of vertigo and nausea. In this way, the research team can correlate measurements of balance and vertigo with objective data about brain activity.

At the time this book is being written, the research is ongoing so there is no neat ending to this story. But the hope is that in a few years, people like Helen will have more and perhaps better treatment options for their PPPD than are currently available.

---

*As Dr. Whitney says, it's like in the movie *Star Wars* when they jump to lightspeed. It's also like the old Microsoft screensaver of moving dots of light on a black background, only here it is big and surrounds you to the edges of your peripheral vision.

7

Sound

Infra and Otherwise

BY NOW, YOU'VE PROBABLY SURMISED that I have a sensitivity to just about anything that can make a person nauseated. And I happened to be at the right place at the right time to experience the rare, sick-making effects of infrasound. Not everyone has had this unique opportunity, and most people aren't susceptible. So it's a dubious privilege, on par with being the only person in the office to get sick with the flu . . . twice in the same season.

First, a little background: What we perceive as sound is actually changes in air pressure, which moves in repeated patterns (what we call sound waves), which are picked up by the inner ear and decoded as sound in the brain. The longer the wavelength, the lower the sound we hear. Generally, human beings can hear sound between 20 Hz and 20,000 Hz (or 20 kHz).* (Hz stands for "hertz," which is a unit

---

*To break that down further, only kids can hear up to 20,000 Hz. Middle-aged folks are lucky to hit a high of 12,000 Hz. The audiologist's rule of thumb is that you lose one Hz cycle for every day you live. For example, if you have the ability to hear 15,650 Hz in 2017, you'll only be able to hear up to 12,000 Hz by 2027. As one audiologist friend said, "Scary!" Yup.

of frequency. Something with a low hertz happens less often than something with a high hertz. With sound, the higher the hertz, the higher the sound we perceive.) Compare this with the audible range of dogs, which is approximately between 67 Hz and 45,000 Hz.* This means that humans have better hearing on the low end of the spectrum, whereas dogs hear a whole lot of very high sounds that don't even register with us. Dog whistles create sound at that high, ultrasonic end—we can't hear it, but dogs and other animals can. Cats hear even better than dogs, with a range of 45 Hz to 64,000 Hz. Elephants and ferrets (yes, ferrets) can hear frequencies as low as 16 Hz, while porpoises can hear frequencies as high as 150,000 Hz.

Here's the interesting bit: Just because we can't hear a sound at a particular very low or very high frequency doesn't mean the sound waves don't affect us. In particular, frequencies below our audible range—anything below 20 Hz—have the potential to cause a number of symptoms, including (and most important for our discussion) dizziness, nausea, seasickness, vertigo, and unsteadiness.

Which brings me to my infrasound experience.

Back in the late 1970s and early 1980s (before DVDs, Blu-rays, and Internet streaming dominated movie entertainment), movie theaters experimented with souped-up sound systems. The most popular was Sensurround, which was first used by Universal Studios to boost the experience of seeing the movie *Earthquake*. The system used extended-range bass to generate infrasound, which the movie-goer could feel but not hear, with internal "tremors" mimicking the action on the screen. The initial box office success of *Earthquake* led to competing systems, including Warner Bros.' version called Megasound.

---

*Interestingly, different breeds have different hearing ranges. One study found that a poodle could hear frequencies as low as 40 Hz.

While supersound systems made movies more of a full-body experience, the buildings felt it, too. Several theaters reported structural damage from the intense infrasound vibrations, with falling ceiling plaster and wall cracks. Most people felt generally shaken, but in a good way. But for a small percentage—like us sensitive folk—the infrasounds induced nausea, vomiting, and a general motion sickness. It would be nice to think that production companies stopped making Sensurround and Megasound movies because they felt bad for those who were made sick, but the truth is that the sound systems took up too much room in theaters, cutting down on the number of tickets that could be sold. Plus, in Multiplex theaters, moviegoers in adjoining rooms also felt the shake. Imagine trying to get a good cry with *The Notebook* while being rocked by a simulated earthquake or alien invasion.

In 1980 I saw *Altered States* at a Megasound theater in Toronto. The movie, which stars William Hurt and Blair Brown, tells the story of a Harvard professor who regressed to a primitive mental and physical state through hallucinogenic drugs and an isolation tank.* Toward the end of the movie, there is a powerful scene in which Hurt's character is struggling down a hallway, banging his fists on the walls to fight back the biological devolution that threatened to take over his body and turn him into a more primitive

---

*While isolation tanks have gone the way of Rubik's cubes and bean bag chairs (that is to say, still around but far diminished in popularity) they were part of popular culture way back then. The tanks were designed to remove as many sensory inputs as possible. They looked like large, enclosed, pod-like bathtubs. A person would sit, naked except for earplugs, in body-temperature water to eliminate most skin sensation and sound. The water would also have high salinity to promote floating. The access doors would be closed, and the interior was dark, eliminating sight. The mythology was that without sensory input, the brain would be free to be more creative, and we could discover deeper psychological truths about ourselves. I never tried one, but I would have given up my collection of fedoras and all my spider plants (many, many spider plants and spider babies) for the experience.

human state, closer to ape than man.* There seemed to be a low rumbling throughout the scene, and as the actor slammed the walls, there was a visceral burst from the speakers. Before he reached the end of the hallway, I had to run—literally sprint—out of the theater, and I promptly hurled into a trash can. My date followed me, just in time to see me bent over with nausea. With lingering illness, I had to go home. I didn't see the end of the movie.

I didn't feel otherwise ill, and because the nausea and general instability came on so suddenly, I reasoned that the movie was so awesomely intense that my sensitive soul couldn't handle it.† A few days later, I went back to the theater (alone this time—I wasn't about to be embarrassed again). I knew what to expect, and I was anxious to see how the movie ended. Again, in the exact same place in the movie, with William Hurt banging down the hallway, I felt sick and had to run out of the theater. I threw up again in the exact same trash can. I didn't see the movie again for another thirty years, when it was showing on television. When the hallway scene came up, I steeled myself, but . . . nothing. Not only was there no nausea, but also I didn't find the scene all that emotional. That's when I learned about infrasound. My emotions—and my stomach—had been manipulated by a couple extra sets of sound speakers.

The Resonance of Your Gut

Why would sound create nausea? There are a few theories, and one—the theory of resonance—has had the distinction of being the subject of both US military research and an episode of *South Park*. More about that in a moment.

---

*It makes more sense when you watch the movie . . . sort of.

†This is exactly the reaction the movie production company wanted to create, that intense, emotional, slightly agitated feeling. At the time, they didn't understand the deeper physiologic effects.

It's a law of nature that everything vibrates. Beginning from electron vibrations at the subatomic level, everything in the world has vibrational energy. When vibration meets vibration, the different frequencies can affect each other. One of the classic ways of illustrating this is to take the example of a child on a swing. When the swing is moving, it has a particular rhythm. If you push the swing with the same rhythm or pattern that the swing is already moving in, you can make the swing go higher. When rhythm matches rhythm, the movement is enhanced. This effect is called resonance. However, if you try to push *against* the cycle, the swing will slow down and all movement could stop. (And the child will be very angry.)

When physics says that everything vibrates, it means *everything*. Including parts of your body. Different body parts have different frequencies.

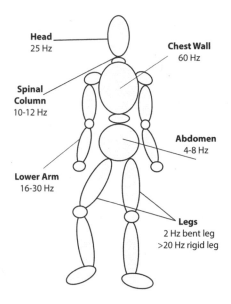

Model for the general frequencies that affect the human body.
Based on information and diagram in Human Body Vibration Exposure and Its Measurement *(Rasmussen 1983)*

Note that many of the body-part frequencies are infrasonic—below the human audible threshold of 20 Hz. When an external vibrational frequency matches a human body vibrational frequency, when rhythm matches rhythm, there is resonance, and the human body will feel something. This is a very simplified explanation of a complex process. Because the body is not a solid structure but is, really, constantly changing, there is no single, perfect resonant frequency. This means that it's harder to predict exactly which frequency will cause an effect on any given part of the body. Plus, any physical effects will vary from person to person or from exposure to exposure. As with seasickness, some people won't experience any symptoms, while others—like me—will become temporarily incapacitated.

Infrasound abounds in our world, from both natural and man-made sources. Perhaps the steadiest form of infrasound in nature is the crashing of ocean waves on the shore—that's why being by the surf makes some people sick. Storms, tornados, earthquakes, avalanches, and volcanic eruptions all create infrasound. The 1883 eruption of the Krakatoa volcano in Indonesia was powerful and loud. It has been estimated that being near the Krakatoa explosion was like standing on a rocket launching pad with no ear protection, and the concussion caused permanent hearing loss in people up to one hundred miles away.* The infrasonic sound waves from that one event circled the globe seven times. Beyond the potential for causing devastating injury and damage, infrasound from such powerful weather events can also cause feelings of fear and dread, motion sickness, and imbalance.†

---

*While infrasound cannot be heard by humans, the sound waves can still damage the ear the way very loud audible sound can. Both audible and infrasound contributed to the injuries caused by Krakatoa.

†With earthquakes, there's the added problem of uncontrollable motion. Between the earth

You know the rumor that animals can predict storms and earthquakes? Turns out it's true, thanks to infrasound. Many types of animals, including cats and dogs, become agitated before severe weather and earthquakes, mainly because of their sensitivity to sounds below the human audible range. Some of the best disaster predictors are elephants and ferrets, two species that can hear frequencies below 20 Hz. Eyewitnesses reported seeing elephants running inland about an hour before the 2004 tsunami in Phuket, Thailand—they were not among the fatalities. There have also been reports of goldfish—which can hear down to about 20 Hz—swimming erratically in their bowls or even leaping from the water before earthquakes.* And the land animal with perhaps the best infrasonic hearing is the pigeon,† which has been known to hear distant storms and to evacuate an area before bad weather hits.

Man-made sources of infrasound include aircraft, rocket launches, explosions, and basically anything with a large engine. Riding in a car, bus, train, helicopter, or ship can potentially cause infrasound-related sickness, including the sensation of sleepiness or fatigue. In fact some experts believe that infrasound is a major contributor to feelings of seasickness and other forms of motion sickness. Even if you're not the one in a car or truck, the road noise from vehicles on nearby highways is a huge source of infrasound. All those low-frequency sound waves can cause quite a lot of discomfort.

---

shaking and the waves of infrasound, earthquakes can create motion-type sickness in up to 30 percent of people who live in quake-prone areas. In Japan, where earthquakes are a regular occurrence, doctors refer to the condition as *jishin-yoi*, which translates to "earthquake drunk."

*Members of the Stanford crew team reported seeing fish jump from the water during the Loma Prieta earthquake in Northern California in 1989.

†While the exact hearing range of pigeons hasn't yet been determined, some sources say they can hear down to 5 Hz

Which brings us to the military research. Human body resonance is of considerable importance to the navy, air force, and
NASA space program because the mechanical shaking, rattling,
buzzing, pulsating, infrasonic vibrations caused by engines in large
ships, planes, and rockets could cause a devil's laundry list of symptoms, including headache, nausea, vomiting, vertigo, dizziness, disorientation, fatigue, irritability, and a general uneasiness. That's
why millions of dollars have been spent to analyze the effects of
vibrations on human health and job performance—it wouldn't do
to have troops getting sick while doing their jobs. In some cases,
the mechanical operations of vehicles have been changed in order
to make troops more comfortable. After all, comfortable military
personnel are better able to perform at peak effectiveness.*

Infrasound can be so potentially devastating that the US government was researching the possibility of an infrasonic weapon—also
called a long-range acoustic device (LRAD)—that could incapaci-
tate or even kill an enemy. Currently, LRADs within audible range
are being used as nonlethal weapons by the US military, by large
ships to defend against pirates, and by police for riot control. It
is basically a very loud noise that creates ear pain and permanent
hearing damage, with no loss of life. According to one article,
victims have dubbed it "The Scream."

Knowing that sonic weapons are possible, and that infrasound
isn't audible but still can have effects on human physiology, *and* that
different parts of the human body have different resonant frequencies, some folks have theorized that it might be possible to build an

---

*Not all human problems from mechanical/engine/vehicle movements are due to infra-
sound—the shaking component alone has the ability to cause physiologic effects. We all
know that shaking a baby can cause irreversible brain damage. It's not that far a leap to
imagine a rocket powerful enough to reach the moon but also powerful enough to shake
its cargo of astronauts to death.

infrasonic weapon that, at the right frequency and volume, could make victims toss their cookies. Or make heads explode. Or loosen bowels on a mass scale. I couldn't find any documented instances of this type of skull-shrapnel-, vomit-, poop-inducing weapon being built and used,* but some people still worry about the possibility. So the dream of being able to use sound waves to make the enemy poop their pants still lies firmly in the realm of fiction. But that fantasy was powerful enough to evoke an episode of *South Park* from the genius minds of creators Matt Stone and Trey Parker.

South Park is an animated show that follows the edgy exploits of a bunch of elementary school kids. In the episode titled "World Wide Recorder Concert,"† Eric Cartman discovers the mythological "brown note," a sound that makes people involuntarily lose control of their bowels. On the show, they find the note (which they pitched as a low bass note, rather than an inaudible infrasound) with the help of musician Kenny G. Of course, in the hands of a group of third graders, the power of the brown note is used for laughs and retribution. Only accidentally do all four million kids participating in the World Wide Recorder Concert blow the brown note at the same time, causing everyone on earth to crap their pants‡ at the exact same time, creating universal devastation. Thankfully, it was just a silly animation.

Blowing [Chunks] in the Wind?

The most controversial and contentious subject in the realm of infrasound is wind turbine syndrome (WTS). Wind turbines are

---

*Seth S. Horowitz wrote a great summary of the topic in his book *The Universal Sense: How Hearing Shapes the Mind*, in which he explored whether these types of physiology-affecting infrasonic weapons are possible. He concluded they are not.

†Season 3, episode 17. And yes, I watch *South Park* regularly, not just as sanctioned "research" for this book.

‡Those are the words used on the show, not gratuitous potty mouth from this writer.

essentially giant windmills that convert the wind's kinetic energy into electrical energy. In a world rightfully obsessed with finding alternatives to fossil fuel energy, harnessing wind power would seem to be nothing but good, but some people who live near wind farms would say otherwise. The turbines create infrasound. And infrasound makes some people sick.

Wind turbine syndrome is a hot-button issue that kindles passionate, forceful arguments on all sides. The disease was "discovered" and defined by Dr. Nina Pierpont, although her work was done on birds, not humans.* With the birth of this disorder, interested people broke into three factions:

- people who report being affected by wind turbine syndrome;
- people who work for "big wind"—those who make money from wind power; and
- scientists, who try to rationally analyze the issue.

While Dr. Pierpont didn't base her book on human research, she knows a lot of people who became ill from wind turbines. She calls the disorder "wind energy's dirty little secret." When wind turbines were erected near her home, she began to hear reports from neighbors of a clear cluster of symptoms: migraine; motion sickness; vertigo; noise, visual, and gastrointestinal sensitivity; and anxiety. (Other researchers also list sleep disturbance, rapid heart rate, visual blurring, panic attacks, and ringing in the ears as WTS symptoms.) Dr. Pierpont began a case series medical study of the disorder and discovered that wind turbines turned homes into nightmarish prisons—many people couldn't sell their noise-slammed homes due to the wind turbines. As Dr. Pierpont puts it, the homes became "acoustically toxic."

---

*She wrote and self-published a book on the topic called *Wind Turbine Syndrome: A Report on a Natural Experiment.*

In some cases, families fled, abandoning their homes to regain their health.

If you visit a wind farm, you can stand and marvel at the wonder of hundreds of enormous turbines all revolving at the same time, and you may not experience illness.* But if you spend a significant amount of time there, as those who live near wind farms do, you'll start to notice at least some effects of being bombarded by audible noise and infrasound all day and all night. If you are among the unlucky few who are sensitive to the sounds, your home may begin to feel like a torture chamber. Check out this description about wind turbines from the diary of one man affected by WTS:

> They are loud. They've been compared to jet engines. A plane that will not take off. There is no gentle swoosh; it is a whoosh noise. They grind, they bang, they creak. The noise is like surround sound, it's omni-directional. It feels like there's this evil thing hovering above you and it follows you everywhere. It will not leave you alone. This noise will not allow you to have your own thoughts, the body cannot adapt, it's a violation of your body. It is a noise that the human body cannot adapt to even after more than a year of exposure. As time progresses the noise becomes even more unbearable.†

To understand the scope of the problem, begin with imagining the sheer size of a single wind turbine. Industrial models, which are the ones creating the most complaints of sickness, stand 328 feet tall. That's nearly 23 feet taller than the Statue of Liberty (305 feet,

---

*Unless you have visual vertigo. Then just looking at the windmills, the rotating motion of the blades, will likely make you feel quite sick.

†www.windaction.org/posts/16670-the-d-entremont-s-letter-filed-with-calumet-county -wi#.V36nY1ePmNg.

1 inch) and ten times taller than an average two-story house. Then take a look at Altamont Pass in California. At its peak, there were six thousand wind turbines running in the area. That's thousands and thousands of structures larger than the Statue of Liberty, all turning and rumbling at the same time. That particular wind farm area is getting smaller. By 2015 the number had been reduced to three thousand turbines. By 2018 all the old turbines need to be stopped, in accordance with advocacy efforts from the Audubon Society,* but they may possibly be replaced by newer models that are safer—for birds, that is. How safe they will be for people who claim to have WTS is debatable.

The problem faced by people with WTS is that it is not a universal ailment—only a small group of sensitive people are affected. That makes it difficult for scientists to devise a research study that can define and replicate the syndrome. Research is made more challenging by the fact that, even among sensitive individuals, symptoms can vary. One person might have sleep disturbance and migraines; another might have ringing in the ears and dizziness; a third person might have nausea and memory troubles. It's nearly impossible to reliably study a condition that doesn't have a definitive set of symptoms. Worse still, those who are affected by wind turbines may also have other vestibular disorders, which would complicate the medical landscape. Where does visual vertigo or BPPV stop and wind turbine syndrome begin?

---

*The changes to turbine laws were spurred by advocacy efforts from the Audubon Society, primarily the Golden Gate chapter. The out-of-date turbine models had been dangerous to many species of birds but especially raptors, which are genetically programmed to focus so fervently on their prey that they don't see the spinning blades of the turbines. The wind turbines killed sixty-seven golden eagles between 2004 and 2014. But according to an article on SFGate.com, "eagles aren't the only flying predators that get nailed by windmills. Burrowing owls, red-tailed hawks, American kestrels and bats are also killed."

Researchers acknowledge that some people living near wind farms have vestibular symptoms, but the exact cause of the illness cannot be definitively tied to the turbines. A review article published in 2011 concluded that current research cannot point to a direct causal link between wind turbines and physiologic symptoms. The authors admit that "wind turbines can be a source of annoyance for some people," a statement that is likely considered a source of annoyance by anyone who feels legitimately ill. Although not explicitly stated, the authors intimate that any perceived health effects are all in a person's head, that people who experience WTS symptoms are so annoyed and stressed that they are making themselves sick. I imagine that sufferers would reply that yes, they are annoyed and stressed, mainly because they feel sick and no one believes them!* Other reviews published in 2014 and 2015 found that exposure to wind turbines increases both "annoyance" and sleep disturbance in a dose-response relationship. That means the closer you are to the wind turbines, or the more time you spend near them, the worse your symptoms will be. That's a better start to understanding possible health effects of WTS—if wind turbines disturb sleep, then it is just a short scientific leap away to say the turbines could cause serious disease.† After all, sleep disturbance is associated with an increased risk of diabetes, heart attack, stroke,

---

*This type of argument was also used for Lyme disease back before the 1970s, when the disease was officially recognized as legitimate, caused by tick-borne bacteria. Before then, doctors thought the symptoms were psychological in origin, and sufferers went untreated. It wasn't until rogue patient advocates took action by conducting their own research and contacting scientists that doctors thought to look more closely at the puzzling condition. So if I sound like I'm on the side of WTS sufferers, it's mainly because medicine has marginalized patients before. I'm all for wind energy, but not at the expense of the health of a significant portion of the population.

†To be clear, not every study has made the link between noise and sleep disturbance. But the 2014 and 2015 studies were reviews, which examined the results from a large number of other studies.

obesity, depression, and high blood pressure. If noisy machinery put my health at serious risk, I, too, would be annoyed.

.   Of course, it is possible that infrasonic noise is affecting everyone—not just sensitive folks—at a deeper level. Deeper than people recognize, and deeper than scientific research can examine in humans. Some animal research has shown that wind turbine noise and infrasound have significant physiologic effects. A Polish study published in 2013 took health measurements of five-week-old domestic geese that grew up within either 50 meters of a wind turbine (that's about 55 yards, so let's call this the close group) or 500 meters (that's about 547 yards, so this is the far group). The scientists looked at weight gain and concentrations of the hormone cortisol in the blood, a reliable measure of stress in the body. A second set of measurements was taken twelve weeks later. The researchers found that geese from the close group gained less weight and had more cortisol in their blood compared with geese from the far group. In addition, geese from the close group generally had less activity and had a "disturbing"* habit of remaining in a tight group. Normal, healthy geese roam freely and explore their environment, but these noise-challenged geese huddled together. One of the overall conclusions by the authors was that "infrasound noise may be a very serious source of stress."

Another animal study, this one published in 2016 by scientists from Great Britain, looked at badgers—short-legged, fat-bodied little creatures related to otters and weasels. The researchers compared cortisol levels in hair from badgers living about 1 kilometer (about 0.6 miles) away from a wind farm and from badgers living about 10 kilometers (a little more than 6 miles) away. They found that the hair cortisol level (a measure of stress) was 264 percent

---

*Disturbing was the authors' word.

higher in the badgers living close to wind farms compared with those living far away. The major conclusion, according to the authors, was that "the higher cortisol levels in affected badgers is caused by the turbines' sound and that these high levels may affect badgers' immune systems, which could result in increased risk of infection and disease."

So what does that mean for people? Is WTS possible? Sure. But as of now, researchers have not yet found a way to measure all the different ways infrasound can affect the body in vulnerable individuals. Companies are using that lack of evidence as an argument for building more wind farms near communities. Basically, the only option for people with WTS is to move, shut down the wind farm, or become activists for change. Sadly, if you want to shut down a wind farm, you'll have better luck pointing to dead birds than to your sick family.

Back to the Movies

Let's return to my movie experience, described at the beginning of this chapter. What happened to make me so sick? When William Hurt was fighting his way down the hallway, infrasonic sound effects punctuated each instance of his fist hitting the wall. The Megasound system used in the theater was capable of hitting a low frequency of 16 Hz, into infrasonic territory. And lucky me—there were only four movies ever made with Megasound.* And full sound effects were released in a very limited number of theaters. Which just goes to show that if there is a chance for an experience to create nausea and vomiting, I'll probably be there.†

---

*The other three were *Outland* (1981), *Superman II* (1981), and *Wolfen* (1981).

†I don't know if this is true, but there's a story about how Walt Disney and several artists screened some animation that had a sound effect that had been accidentally slowed from 60 Hz to 12 Hz. When played through the theater speakers, everyone in the room became

Can You Hear Where Your Feet Are?

Infrasound is an example of how sound might hurt our sense of balance. But researchers are currently working on defining and perfecting ways for sound to help us stay on our feet. To get that side of the story, I contacted Dr. Timothy Hullar, professor of otolaryngology at Oregon Health & Science University. Dr. Hullar is the quintessential cool doc. His manner is laid-back, and his free-form conversational style allowed our discussion to roam from wind turbines to hearing aids to medical marijuana.

Most of us don't appreciate how much our hearing contributes to our sense of balance—we use the sounds that swirl around us all day long (what scientists call environmental cues) as aural landmarks. Just like visual landmarks, these sounds help us determine which way is up, down, forward, backward, left, and right. When you open your eyes and see that maybe you're tilting a little bit, you adjust based on those visual cues. That's obvious to just about everyone. The connection to the inner ear isn't quite as obvious, but if you can localize a sound source in the world—pinpoint where it's coming from—that's a valuable bit of information. So when you "open your ears," it can be just as powerful a balance tool as opening your eyes. Plus, sound provides passive, twenty-four-hour-a-day cues.

"The auditory environment presents itself to you from all dimensions and directions," says Dr. Hullar. "Vision, if you're not actively looking at something or seeing it out of the corner of your eye, you're not seeing it. With audition, if something is happening behind you, you can still use that information and you're still aware of it."

---

nauseated and dizzy, and the feeling of illness lasted for several days. My emails seeking confirmation have gone unanswered.

That realization led Dr. Hullar and his colleagues to look more closely* at the connection between ears and balance. Their's was the first lab in the world to show that when people have hearing loss, giving better hearing cues can improve balance. What that means is simple yet revolutionary: Balance improves when people with hearing loss use hearing aids. As Dr. Hullar puts it, "Hearing aids are also balance aids."

This finding is important for all of us. Researchers from Finland tracked the lives of average citizens and found that people with hearing loss tend to fall.† This was groundbreaking. We tend to think that both hearing loss and falling are merely casualties of old age. Now we know that there is a closer relationship, that loss of hearing causes some loss of balance. The next logical step is for scientists to investigate how people use sounds for navigation—what frequencies, loudness, and quality of sounds could be used to make home environments safer for everyone as we get older and lose some hearing.

The Finnish researchers gathered data using an interesting method. Subjects stood on a platform with four pressure sensors beneath it, which allowed researchers to measure sway—when you're standing, no matter how solid you think you are on your feet, you will always sway, even if just a little. Your body is constantly making little adjustments to keep balance—maybe a little more weight on the left side or on the balls of your feet—and as soon as you make one adjustment, your body will sway and you'll have to make more adjustments. It turns out that sway is a measure of balance ability, with ratings that vary from "rock-solid stable" to "can't even stand up." There's a natural decline in balance with age,

---

*Perhaps "listen more closely" would be a better phrase here.
†This research was later repeated here in the United States.

and of course there's a decline with people who have other sensory problems, such as vestibular or proprioceptive losses. By using the pressure platform, researchers were able to measure gradations in changes in an individual's sway.

In one study, people stood on the platform in the dark with their eyes closed. The amount of sway was measured in silence and also with sound—a white noise—projected from four speakers around the subjects' heads. Balance was dramatically improved when sound was introduced. And in a uniquely optimistic finding (for scientific research), the experts found that the worse off a person was in terms of balance, the greater the improvement they had with the addition of sound. Then the researchers went a little further and changed the way the platform worked. This time, when a person swayed as a way to keep balance, the platform swayed, too,* removing somatic and proprioceptive cues and making it much more difficult to balance. In the dark with no sound, many people couldn't stand up. When the sound was turned on, subjects *were* able to stand up. So dramatic! With the addition of sound, people went from falling to standing. Imagine what that means for the frail elderly, whose lives are at risk with every fall.

But that's just about standing. What about walking—balance in motion? Dr. Hullar has a study for that, too. Take the basic premise of marching in place in a darkened room, with the goal of maintaining the same orientation such that if you start the experiment facing the north wall, you end the experiment facing the north wall. After a few minutes, if you have good balance, you won't rotate much—you'll still be facing that north wall. But if your sense of balance is off, you can start out facing north but end up facing south. The researchers had people perform this task in

---

*This is called sway referencing.

silence, then again with white noise. People rotate less when noise is present.* This study went a step further and demonstrated that it also matters where the sound is coming from. Sound that comes from behind you isn't as helpful as sound in front of you. I asked how this might be helpful in real life, and it turns out this scenario is not so far-fetched. Imagine trying to walk through your home in the dark—you'll be more successful at finding the light switch (and avoiding the furniture) if there is a radio turned on across the room or if someone talks to you while you're walking. Just a helpful tip to remember the next time the power goes out.

Helping Future Olympians (Maybe)

Sometimes life provides its own unintentional research subjects. Physicians who treat people with ear problems, such as children with otitis media with effusion (OME), find themselves witness to some amazing sound-balance demonstrations. We all have fluid in our ears, and most of the time it drains out through the Eustachian tube, which connects the inside of the ear to the back of the throat. If the Eustachian tube becomes blocked, fluid can build up in the ear, resulting in the condition called OME. Of course, bacteria thrive in stagnant fluid, so ear infections often go hand in hand with OME. This condition is most often seen in young children because their Eutstachian tubes aren't fully developed—they are shorter, straighter, and floppier than the tubes of older children and adults. This makes it easier for the tubes to become blocked and

---

*As Dr. Hullar described this experiment, I wondered what sophisticated equipment was required. I imagined a large room with a rotating platform connected to a computer that allowed the researchers to get step-by-step measurements of a person's rotation. So I had to ask. Dr. Hullar laughed. The total mechanics involved two students, a protractor, and a radio in an otherwise empty room. Science often has to make do with the bare minimum.

for bacteria to enter. Those two conditions mean that some young children have frequent ear infections. But even without bacterial infection, the fluid in the tubes leads to muffled hearing, a sensation a little like putting cotton balls in your ears. This condition can remain for several months until the child outgrows the condition.

The problem is that OME hits at a critical time in children's development. They are often acquiring speech and language skills, which can be hampered by hearing loss. Physical development may also be affected. One of the treatments for OME is myringotomy—a surgical procedure that puts a small tube in the eardrum to drain fluid and relieve pressure in the ear. Experts who treat OME report that kids who get the tube surgery have much improved balance. It makes sense. The kids are in the middle of developing their motor skills and have huge leaps in social and intellectual learning and cerebellar development. If they can't hear, they will be held back in their development. But with the tubes—well, doctors tell almost identical stories. They see it all the time: A mom and dad come to the office for a follow-up visit after tube surgery and say, "My child was really clumsy until the tubes went in." After surgery they're like little Olympians. Experts see it and believe it. Unfortunately there is no definitive supporting research, so most insurance companies* still don't support tube surgery for most cases of OME. Still, doctors see a big jump in achieving motor milestones with

---

*This includes the Oregon Health Plan, the public insurance system for Oregonians. Even though Oregon can boast of having some of the top researchers and practicing audiologists in the world—including Dr. Timothy Hullar—the state does not provide coverage for tube surgery for children with chronic otitis media with effusion (OME) for a sole indication of hearing loss. As of 2017, this surgery—which improves balance and allows affected children a chance to reach their full physical and social potential—ranks below hemorrhoids on the list of Oregon's priorities. Seriously. Sagging eyelid surgery (blepharoplasty) is covered, but tubes for OME is not. Doctors there (and everywhere) would be thrilled if all their young patients could take advantage of the remarkable gains seen with this balance surgery.

these kids. They're hearing the environment around them, paying attention to those audio landmarks, and balancing like gymnasts.

The same type of dramatic improvement in balance can happen with adults with hearing loss. Dr. Hullar told me about a woman he knows who is very severely deaf. She knows that her balance is bad when she takes both her hearing aids off, but balance is even worse when she takes just one of them off. That's because she's unbalanced in her hearing—almost worse than no hearing is having unbalanced or contradictory hearing inputs. This is a common risk among people who need hearing aids, who often don't understand how much sound contributes to balance. Many people with hearing deficit in both ears opt to buy just one hearing aid instead of two, sometimes for vanity's sake, sometimes for financial reasons. But all signs point to two hearing aids as significantly better than one. With two, you can spatially localize sound. It's like the difference between having vision in one eye or two—you will see the same things, but having two eyes gives you depth perception. Hearing with two ears gives you the auditory equivalent, a kind of sonic depth perception.

like this: I call police department A, who says that the goggles are available at police department B, but they tell me that their goggles were broken so try police department C, but their goggles are lost so call police department D, but if that doesn't work then try the state highway safety commission. It was a months-long, extended version of telephone tag. After a while, it felt like I was hunting for something elusive, like Bigfoot, and proof of existence would be just around the corner.*

Finally I found them in Garner, North Carolina. With a current population of less than thirty thousand, Garner is growing in all the right ways. It boasts a performing arts center, 350 acres of maintained parkland, and a Fourth of July celebration that includes a performance by the North Carolina Symphony. It could serve as a quaint small town in the movies. In 2013 the National Civic League named Garner an All-America City. My contact, Sgt. Chris Adams, leads the section of the Garner Police Department that handles drunk drivers, car wrecks, and other traffic safety issues.

Before our meeting, I tried to imagine what a traffic safety supervisor might look like. Except for the hair—short and dark, which I imagined blond—Sergeant Adams hit all the points. He is of about average height, with a fit build and a more casual stance than you might expect from a longtime member of the police department. He is friendly, with a smile that imparts genuine cheerfulness and a magnetic air that makes every sentence he utters seem significant, even when he was just introducing me to the department's goldfish.†

---

*My biggest lesson was that it is really easy to lose or break the goggles. This is based solely on what I was told when I tried to find a test pair and is not a commentary on the goggles themselves, which seem very sturdy.

†The department has a medium-sized tank sparsely decorated with blue and white gravel, three police cruiser toys (one crashed and half submersed in the gravel), and something fuzzy that looks like a rabbit's foot. There are also two goldfish. The older one is about

Being a small town doesn't make Garner immune to the wrong side of alcohol. According to Sergeant Adams, officers can start cruising the streets any night of the week and be guaranteed to pull in an impaired driver. "My mother always told me, 'Nothing good ever happens after midnight,'" Sergeant Adams said. "She was right. And I would add 'Don't be out on the roads after 2 AM.' That's when the bars close. That's when things get really bad."

We went to the Garner police muster room, where Sergeant Adams had a large briefcase with the drunk glasses. They look like any pair of full-protection eye safety goggles but a little larger and with thick lenses. The Coke-bottle glasses of the goggle world. Some lenses are clear, and some are tinted dark gray to simulate being drunk at night.

Fatal Vision goggles are available in five impairment levels that correlate with different blood alcohol concentration (BAC) levels:

- BAC of less than .06
- BAC of .07 to .10
- BAC of .12 to .15
- BAC of .17 to .20
- BAC of .25

In all fifty states, the legal blood alcohol concentration limit for driving under the influence (DUI) or driving while impaired (DWI) is .08. That means the lowest-powered goggles corresponded to a level of "I had two drinks but I'm not drunk." I skipped that one. In all, I tried three levels of goggles, and one nighttime pair. I started with the lowest legally drunk level of BAC .07 to .10.

---

seven inches long and is named Twenty-Eight Cents, in honor of how much he cost when he was purchased. Twenty-Eight Cents has a new partner (I love that they used police lingo for the fish), a three-incher named Seven Ninety-Nine. Guess why.

I put them on over my own glasses. It was nothing like the more subtle effect I had expected. Several months earlier, I had spoken with Deb Kusmec, chief operating officer of Innocorp, the manufacturer of Fatal Vision goggles. She told me that the drunk effect is caused by optical prisms that alter your vision—the lenses shift everything slightly to the right. I thought that information would prepare me for the effect of the goggles. But even knowing the mechanism behind the drunk effect, I still got the full experience.

Within seconds of putting the goggles on, I could have believed I was drunk. Nothing looked "shifted right." Instead there was a distortion that made the floor seem a little farther away and had me walking into tables in the muster room. I couldn't walk straight without putting my arms out for balance. Even standing still, I felt like the room was moving a little. It was probably due to the movement of my eyes struggling to make sense of the prism view, but it was powerful. I was dizzy, with artificial vertigo. I really felt drunk. One part of the demonstration included putting me through the typical field sobriety test. I couldn't make it past the first part of the first task—the "walk a straight line" test. The instructions were to look at my feet, walk nine steps heel-to-toe, with hands down at my side. So . . . step one: look at my feet. That's as far as I got. The simple act of looking at my feet gave me the most peculiar feeling because my proprioceptive cues were mismatched to my vision cues—my body told me where my feet *should* be (proprioception), but the goggles showed my feet shifted to the right (vision). That small disconnect between the two balance senses set off my sensitivity to motion sickness. My head was spinning, and nausea kicked in. I couldn't take a single heel-to-toe step without raising my arms for balance and teetering sideways. Had this been a real police stop, and had I really been drinking, I absolutely would have been arrested.

Then, Sergeant Adams reminded me that I was wearing only the lowest level of above-legal-limit glasses. There were three more levels to go. I tried two, plus one nighttime simulation pair. With each, my balance was even more compromised. By the last one, I couldn't maneuver between chairs without grabbing on to keep from falling. If this were real-life drunk, I wouldn't be able to pedal a big-wheeled tricycle down a hallway without crashing into a wall. It turned my stomach to think about what might happen if I were ever to get behind the wheel of a car while this impaired. I'm sure most people feel the same way, but only when they are not drunk. The problem is that when a person is drunk for real, judgment is also twisted. Impaired people can't recognize how off balance they really are. It's like how people with Alzheimer's disease can't know what they don't remember, so they deny having memory problems. Alcohol affects perception and judgment so drunk drivers don't recognize they are barely able to walk a straight line, let alone drive a car. But the effects of alcohol on balance (and physical and mental capabilities) are very real.

Like Helium for Your Ears

Fatal Vision goggles just approximate the aftereffects of drinking. But what happens when you really are tipsy? Why does alcohol make you feel, by turns, dizzy, unbalanced, and whirly? I put that question to one of the country's foremost experts on the topic— Kevin Strang, PhD, distinguished faculty associate in neuroscience/ physiology at the University of Wisconsin-Madison.

Dr. Strang's mission is to educate people about effects of alcohol on the body—the molecular-level physiology—so that everyone can make informed decisions about drinking. He's not opposed to alcohol; he just believes that it's important to understand the underlying effects and outcomes. I have heard several of his lectures, and I must say that Dr. Strang has a wonderfully understandable and approachable speaking style, and he gave me a whole new appreciation of what happens when a writer (namely me) walks into a bar.*

When it comes to balance, alcohol is a triple threat—it affects the vestibular system of the inner ear, proprioception, and vision. A dysfunction in any one of those areas is enough to trigger dizziness, but when all three are involved, falling down becomes almost a done deal.

If you recall from chapter 2, a major part of our sense of balance comes from the functioning of the semicircular canals in the inner ear. In each canal is a flower bud–shaped structure called the cupula, which contains tiny hair cells that feed information to the brain. And the tubelike loops of the semicircular canals are filled with a thickish fluid called endolymph, which also surrounds and bathes the cupula.

Under normal (i.e., sober) conditions, that system works perfectly: The endolymph swishes through the semicircular canals in response to rotational movements of the head, so we know where we are in space when we nod or flip or twist or turn a cartwheel. The endolymph moves through the canals and presses on the cupula, bending the tiny hair cells, which signals the brain about our movements. So our brains are kept up to date about where we are, what we are doing, how we are moving . . . and therefore we are able to maintain balance.

But then we go out with friends, have some drinks, and stumble home. Why the stumble?

---

*If you want to see Dr. Strang in action, one of his alcohol lectures to incoming university students is available online from the Video Library of the University of Wisconsin School of Medicine and Public Health: https://videos.med.wisc.edu/videos/3393.

I never really thought about it before, but if you had asked me a few years ago what happens to alcohol in the body, I would have had some vague idea of the alcohol going into the digestive system, then entering the bloodstream, then hitting the brain, and finally being filtered and detoxified by the liver. In actuality, when you drink alcohol it affects every cell in your body . . . and it hits different parts of the body at different rates. Within the semicircular canals, alcohol builds up faster in the cupula than in the surrounding endolymph. And—most importantly for the effect on balance—alcohol is lighter than endolymph. Think about a helium balloon. Helium is lighter than the air in our atmosphere, so the balloon will always go straight up. You can turn the balloon, bat it around, try to hold it down, but it will, eventually, float up. Within a semicircular canal infused with alcohol, there is *more* alcohol in the cupula, making it float up regardless of how you move. But—and this is the critical *but*—the cupula is supposed to move in response to the changing flow of endolymph. It isn't supposed to float against gravity like a helium balloon!

Imagine you are drunk. Unbeknownst to you, your alcohol-soaked cupula is floating upward against gravity within the semicircular canal instead of responding to the endolymph current. That wayward cupula sends a messed-up signal to your brain that the body is moving. That message is constant for as long as the cupula is bent in the wrong direction, so the sensation of movement continues, even while you are standing still or lying down.*

---

*In fact, the erroneous signals are most noticeable when you stop moving, close your eyes, and lie down. With your eyes closed, you don't get the visual input your brain needs to help maintain balance. Once you close your eyes, the bad signals from the floating cupula feel stronger. If you are standing, the more vertical semicircular canals become activated, so you feel like you are tipping to one side, or forward or backward. But when you lie down (or fall down, as the case may be), you activate the semicircular canal responsible for detecting rotation, leading to the infamous "spins" or "whirlies."

This density difference, where the cupula is lighter than the surrounding endolymph, lasts about three to five hours. Then the amount of alcohol is equal in all parts of the inner ear, and the illusion of spinning stops. But brace yourself—that's just the eye of the vertigo storm. As alcohol leaves your body, it leaves the cupula more quickly. Since alcohol is lighter than endolymph, that means that the cupula becomes temporarily heavier than the surrounding fluid. Instead of rising up against gravity, the cupula presses downward, and the illusion of spinning starts all over again—but this time in the opposite direction.

Hey, Bartender, Can You Turn Up the Music?

Alcohol's balance effects go way beyond the inner ear. Your entire brain is temporarily changed after a few drinks. No matter how little you think you are thinking, your brain is constantly idling, even when you're zoning out in front of the television or staring into space. Everything that goes on in the mind—whether you're aware of it or not—happens at the junction where two or more brain cells (neurons) meet, an area call the synapse.*

All communication among neurons happens via electrochemical signaling. Let's start with a thought, like *I'm ready for lunch.* The thought begins with a spark of electricity (what happens when we say a neuron "fires"), which causes an electrical impulse to travel down the body of the nerve cell. The end of the cell is not directly connected to other cells. Instead there is a tiny gap, or synapse. When the electrical impulse reaches the end, the neuron releases a chemical messenger (neurotransmitter) into the gap. If the receiving neuron gets the message to fire, it creates its own

---

*There are about one hundred billion neurons and well over one hundred trillion synapses in the human brain.

electrical impulse, which travels down the cell to release its own chemical messengers, ad infinitum. In this way a single neuron may communicate with about ten thousand other neurons. So the *I'm ready for lunch* thought can become a decision to eat lunch, a plan of what to eat, and actual lunch-eating action, all governed by neurons and their communication with other neurons.

There are dozens of different types of neurotransmitters that can be released into the synapse. But the two important ones for this discussion are glutamate and GABA.* Glutamate is excitatory, which means that it encourages other brain cells to fire, stimulating brain activity. When you feel sharp, like your brain is working well and firing on all cylinders, that's thanks to glutamate. On the other hand, GABA is inhibitory; it dampens brain activity. A little GABA is a good thing—it decreases stress and makes us feel calmer. Without GABA we would feel too wired and overstimulated.

Alcohol increases GABA in the brain. So it makes sense that we initially feel calmer after a little alcohol. But with too much alcohol, all brain function slows down in a dose-dependent effect, such that the more you drink, the slower your thinking. In particular, alcohol depresses two brain functions critical for balance: sensory inputs and motor outputs.

Sensory input refers to how certain sensations from the outside world reach your brain. When you drink alcohol, the first part of the sensory system to be dampened (and eventually cut off) is the eighth cranial nerve. That's the big nerve that carries two streams of information: the vestibular branch, which gives balance signals, and the cochlear branch, which sends signals related to sound and hearing. You can tell that a person is starting to get drunk when

---

*GABA is shorthand for gamma-aminobutyric acid.

she keeps asking you to repeat yourself or when he wants the music turned up—drunk people can literally lose their sense of hearing, albeit temporarily. The balance information from the inner ear is also inhibited. That's why field sobriety tests work as a drunkenness gauge.

Because the tests have been depicted on TV shows galore, both dramas and comedies, probably everyone knows that field sobriety tests include walking heel to toe in a straight line (known officially as the walk-and-turn, or WAT) and the one-leg stand (or OLS).* Although a police officer will be measuring several different aspects of the subject's ability to perform these tests, the key balance portion is engaged by telling the person being tested to look only at his or her feet. By now you can probably guess why. Remember, our ability to balance is affected by three main sensory systems—the vestibular apparatus of the inner ear, proprioception, and vision. Requiring test-takers to look only at their feet narrowly restricts visual input, giving greater weight to the vestibular inputs; but if those inputs are inhibited by alcohol's effect on the eighth cranial nerve, then people lose their balance. Plus, looking down exaggerates the effect of gravity on the helium-filled cupula in the horizontal semicircular canal. The result will be a sensation of spinning, which makes passing this part of the test really difficult. The field sobriety test would be even more effective if people were required to close their eyes, but police departments wouldn't be able to deal

---

*The third part of the field sobriety test is the test for horizontal gaze nystagmus (HGN). The person being tested will be asked to follow the police officer's finger without moving his or her head. The officer will be watching for the person's eyes to quiver horizontally. This nystagmus will occur because the eyes are hardwired to the inputs from the semicircular canals, and the alcohol-dense floating cupula in the semicircular canals is giving signals that you are moving, even when you are not. But your eyes don't know the difference, so they move, too. There's no way to beat this part of the test—it is simply a function of your brain's wiring on alcohol.

with the injuries—take vision away entirely, and a really drunk person will topple over almost immediately.*

As sensory inputs become untrustworthy in our drunk brains,† our ability to stand upright becomes a little off. We become less stable. When asked to stand still, we sway more than usual. And without us realizing it, our posture changes in ways that the brain believes will help keep us from falling down. This was demonstrated most effectively by a group of researchers from Sweden and the United Kingdom who have done extensive work on balance and alcohol consumption. They took a group of twenty-five people and made them tipsy once a week for three weeks. Specifically, the test subjects stood on a platform with 3-D motion-tracking devices on their bodies, which could measure even the tiniest changes in body position. Their posture was measured with blood alcohol concentrations of 0.0 percent (no alcohol), 0.06 percent (just under the legal limit of intoxication in the United States), and 0.10 percent (slightly above the legal limit for intoxication). None of these conditions were enough to be considered highly intoxicated or doing-cartwheels-on-your-neighbor's-lawn drunk. To get to a BAC of 0.10 percent, an average woman would have to drink three glasses of wine, and a two-hundred-pound man would have to drink about five glasses.‡ Not an over-the-top amount. Still, the results of the

---

*Seriously, don't try this at home. A person can fall hard. While it may seem like a fun party game, it's more of an invitation to pay a friend's medical or dental bills.

†No kidding—I originally typed "bunk drains" instead of "drunk brains." It seems like an appropriate spoonerism.

‡For the research, the subjects drank "70% ethanol diluted in elderflower juice." I had never heard of elderflower juice, so I did a little extra research. Elderflower cocktails are popular in many parts of Europe and the United States. I tried to track down elderflower juice and found St-Germain elderflower liqueur. I thought it sounded like a rare and exotic drink, but the guy at the liquor store told me it was pretty popular. For the record, for anyone as out-of-the-loop as I seem to be, I tasted it. (My dedication to complete research knows no bounds.) The St-Germain website claims that the taste is "neither passionfruit nor pear, grapefruit nor lemon." To me it is like nothing else I have tasted, a florally-fruity flavor that is so light it barely rests on the tongue before the taste vanishes.

study showed that when we get drunk, we hold ourselves more rigidly and lock our knees as a kind of protective defense against having the knees buckle. In addition, when drunk we tend to hold our heads differently—a little back and deviated to the left. This is noteworthy because it suggests that our basic perception of up is impaired when we drink alcohol, a major trigger for balance problems.

Here's a fun little side note about GABA and alcohol. You know that annoying person who swears that he is better at playing darts or video games when he is drinking? Well, buy that guy a drink, because he's right. But only up to a point, like, say, a drink or two. GABA inhibition has the effect of blocking out stray thoughts and sensations, helping a person become singularly focused on the one task he is consciously performing. A sober person playing darts might also be thinking about his terrible day at work or the presentation he has to give on Monday, and he may also notice that his friend just walked into the bar and that a song he really likes just came on. A sober person can concentrate, but his focus will be shared by all the normal background thinking that is always going on. After one or two drinks, the increase in GABA in the brain blocks out those extraneous thoughts and sensations. When that happens, darts become essentially the only thing in the world. The slightly inebriated person feels one with the dart and dartboard. The Force is with him.* But the GABA-dampening effect is dose-dependent. With more than a couple of beers, processing time will slow so much that performance gets worse.

You and Your Hundred-Pound Legs

The third part of alcohol's triple threat involves changes to motor output, or how your body responds to commands from your brain.

---

*"The Force" would be a great beer brand name. Of course, you could only really market it to *Star Wars* nerds.

Here's how motor control typically works in a sober person: Imagine that there are two dumbbells in front of you—one weighing five pounds and one twenty pounds. You know from experience approximately how much force you need to pick up and curl those weights. When you decide which one to pick up, the higher centers in your brain say "I'm going to pick up the twenty-pound weight." Then other—let's call them middle—processing centers pick up on that thought and relay the message to your body, saying (in effect), "It's a twenty-pound weight; prepare for heaviness." Those middle processing centers send a cascade of signals down your spinal cord, which relays the signals out to your biceps. When you lift the dumbbell, your mind and body are prepared for the heavier weight, so your muscles exert the correct amount of force to get the job done. That's called matched effort output—you intend to pick up a twenty-pound weight, the weight is indeed twenty pounds, and the muscle output is appropriate for that heavier weight.*

With alcohol, the system gets a little messed up. One of the neurotransmitters in our brains is dopamine, which is a pleasure hormone. It's one of the reasons we get a happy buzz from alcohol. But dopamine also dampens some of the cascade of signals to the body from the brain. More alcohol equals more dampening, or inhibition. What happens, then, is that you look at the twenty-pound dumbbell and say, "I'm going to pick up that heavy dumbbell." But because some of the signals on the way down to the muscles are blocked, you can't rally the appropriate

---

*We have all had the experience of an unmatched output. That happens when we decide to pick up what we think is a heavy object only to have it fly up in your hand because the object was actually much lighter than anticipated. Years ago I had a friend show me a series of marble eggs (quite heavy) and then a blown-glass egg decorated to look marble. She's lucky I didn't accidentally crush it or send it across the room from the excess exertion!

number of muscle cells to move that muscle. Suddenly you can't pick up that dumbbell. That may seem like a "so what" scenario, but it becomes more important with activities we take for granted, reflex activities like walking up stairs. You know how heavy your legs are,* and your brain is accustomed to sending the appropriate signals for you to lift your feet high enough to climb a step. But if you're drunk, some of the signals to the leg are blocked, so instead of lifting your foot nine inches (as you intend), it lifts only seven inches. You'll trip on the stairs and fall on your face. Or you'll run into a wall or stumble over your own feet. When you are drunk, lack of balance is reflected in every movement.

This is similar to what happens to people with multiple sclerosis, which permanently affects how the body receives and responds to signals from the brain. With multiple sclerosis, nerve cells lose their electrical insulator, called the myelin sheath, which normally allows the nerve signal to travel true all the way from beginning to end. Without the sheath, the signals are lost before they get to the end. The appropriate message is sent from the brain, but it doesn't make it all the way to the muscles. A person with multiple sclerosis will form an intention in their brain and say, "I'm going to lift my leg nine inches and step on the stair." But because of the disease, only roughly 70 percent of the brain's signals actually get to the muscles. The result is that, as with alcohol-blocked signals, the leg doesn't lift high enough, resulting in a trip on a staircase.

---

*Just for kicks: A human leg weighs approximately 17.5 percent of total body weight. So if a person weighs 150 pounds, one leg will weigh about 26.25 pounds. The "hundred-pound legs" in this section title was meant to be illustrative and maybe humorous, not accurate. I have not been able to find anyone who knew the increase in perceived weight of a leg per alcohol dose.

From ACE to OTC

Alcohol is just one drug among many that can cause dizziness. Just look at the package inserts or labels for any medication, over-the-counter or prescription, and chances are one of the possible side effects will be dizziness. In fact it's hard to find a drug that couldn't induce dizziness in some people, under certain circumstances. I tried and had no luck. A look in my own medicine cabinet revealed that *everything* there had a warning about possible dizziness, including acetaminophen (Tylenol), ibuprofen (Advil), allergy medications (Claritin and Sudafed), Pepto-Bismol, NyQuil, and Gas-X. Even water can be a risk—overhydration can lead to water intoxication, which has been documented to cause vertigo.*

To learn more about the effects of pharmaceuticals on our sense of balance, I spoke with Dr. Emerald Lin, assistant attending physiatrist at the Hospital for Special Surgery in Manhattan. My main question was why so many medications set us off balance—are we that easy to make dizzy?

"The short answer is yes, it is very easy for drugs of any sort to make a person dizzy," said Dr. Lin. "The long answer is that there are many different processes and inputs that our brains have to put together, and they all can influence balance and dizziness. There are central inputs—vision, the inner ear, and parts of the brain—and then there are proprioceptive pathways. The brain has to process all these inputs . . . so there's a lot of room for error."

---

*Too much water dilutes normal levels of sodium in the blood, wreaking havoc on the electrolytes needed for normal brain function—and life. It can be fatal. Too much salt also causes cells to absorb more water, and they swell. When brain cells swell inside the bony cranium, they malfunction. And if the brain swells inside the cranium, it compresses the entering blood vessels, so the brain can be deprived of oxygen, as well. Water intoxication is very rare. You would have to drink several gallons of water in a day, which is really difficult to do accidentally.

It makes sense. We take medications because we feel something is wrong, so we start out with different baseline levels of unwellness. Whatever caused that initial sickness may be the actual reason we felt dizzy, unrelated to the drugs we take to feel better, or the dizziness may be a result of medications interacting with our own specific and unique physiology.

"And even anxiety can provoke subjective dizziness," said Dr. Lin. "That's tricky!"

The upshot is that dizziness as a side effect is the result of a complex physiologic stew that comes from combining an individual's initial state of health, the reason for taking a medication, the properties of the medication itself, the dosage of medication taken, and the person's psychological state. And then it's a question of how well your brain can adapt to the different chemicals and translate any new signals from the vestibular, visual, and proprioceptive systems. If you feel dizzy or off balance after taking any drug, talk with your doctor, especially if it is a prescription medication. There are often alternatives that might be safer for you, or your doctor may have to adjust the recommended dosage. (Don't stop taking any prescription medication without consulting your doctor.)

That said, there are some groups of medications that are more likely to cause dizziness, vertigo, or loss of balance. At the top of the list are drugs used to lower blood pressure, which include antihypertensives, such as ACE inhibitors and calcium channel blockers, and diuretics. These are considered some of the worst offenders by virtue of sheer numbers: Approximately 30 percent of people in the United States have high blood pressure, and it's a lifelong condition that sometimes requires taking more than one medication to get numbers into the healthy zone. These medications cause dizziness, in part, as a result of doing their job. Every time you stand up, your body has to adjust blood pressure to make sure blood gets pumped all the way up to your brain. Nearly everyone has experienced that lightheaded feeling you get if you "stand up too quickly." It's just your brain needing blood when your body was too sluggish to get it there fast enough. People with naturally low blood pressure have this sensation quite often. Blood pressure medications are known to cause this temporary dizziness, a condition doctors call orthostatic hypotension. The dizziness usually goes away in a few seconds, and your body often adjusts to the medication over time. But some people may become so lightheaded that they faint and fall. That's dangerous and should be reported immediately to your physician.

Other dizzy-making drugs are those that change the quantity or availability of neurotransmitters, the chemical messengers responsible for brain signals. There are six neurotransmitters involved in the feedback loop between the hair cells of the inner ear and cells that contribute to the vestibular ocular reflex, the hardwired eye movements in response to changes in the vestibular system. Some neurotransmitters can drop blood pressure; others work on the nausea centers and can even change our perception of the world. The most neurotransmitter-changing medications are selective serotonin reuptake inhibitors, or SSRIs. These antidepressant and antianxiety drugs are better known by their brand names, which include Prozac, Zoloft, Lexapro, Celexa, and Paxil. What's especially interesting is that SSRIs can cause dizziness and vertigo when you take them *and* when you stop taking them. The latter even has a name—antidepressant discontinuation syndrome, which describes the highly unpleasant constellation of symptoms that can happen when you stop taking an SSRI. The two balance-related symptoms are dizziness, which occurs with even small head movements, and a kind of buzzing or zapping feeling in the brain associated with moving your head or eyes left to right or right to left. (The eyes are involved because of their hardwired connection to the vestibular system.)

Although no one knows exactly why SSRIs affect dizziness, researchers in New Zealand suggest that it has to do with structures in the brain called vestibular nuclei, which are basically relay points between the vestibular nerve and the midbrain and cerebellum, brain centers that help control movement and balance. They postulate that because the vestibular nuclei contain lots of serotonin receptors, we should expect crazy dizziness signals both by making serotonin more available (which happens when you start taking an SSRI antidepressant) and by taking serotonin away (which happens if you stop taking the drug abruptly).

Other drugs that are known to cause dizziness include anti-inflammatory medications (such as ibuprofen, diclofenac, naproxen, acetaminophen, aspirin), cholesterol-lowering drugs (such as simvastatin and atorvastatin), antifungals (such as amphotericin B and fluconazole), antipsychotics (such as chlorpromazine, clozapine, and thioridazine), antiseizure medications (gabapentin), sedatives/hypnotics (such as Valium, Xanax, and Ambien), painkillers (such as hydrocodone and oxycodone), and drugs taken to control Parkinson's disease (bromocriptine and levodopa/carbidopa). And this is not a complete list—these are just the large categories of medications that cause dizziness in a fair number of people who use them. Virtually any drug can cause dizziness in some people.

Perhaps strangest of all, however, are the drugs designed to fight dizziness that can actually *cause* dizziness.

"It's sometimes funny," said Dr. Lin, "because you have medications that are supposed to be used for dizziness, but they can also cause dizziness. In trying to balance the inner and the outer inputs, and the processing, and the compensation, the brain plays a big role in that. It disables the inner ear's reaction so much, to the point where the brain doesn't know what to make of it."

And as we have seen over and over again, when the brain is confused, dizziness can't be far behind. Drugs commonly used to treat vertigo include antihistamines (such as meclizine and even diphenhydramine, the active ingredient in Benadryl), benzodiazepines (such as Xanax and Valium), and anticholinergics (such as scopolamine, which is used as a motion sickness remedy). All these medications work by suppressing the vestibular system, inhibiting vertigo and vestibular-related dizziness, but they can cause medication-related dizziness.

As if causing temporary dizziness weren't bad enough, some drugs do actual damage to the inner ear, leading to temporary or permanent vertigo. These ototoxic drugs can damage the cochlea (the portion of the inner ear responsible for hearing), the tiny hair cells in the inner ear, or even the large vestibular nerve. Taking them may result in loss of hearing, permanent vertigo, or both. So why would people risk taking these drugs? Mainly because they may not have a choice. For example, many chemotherapy drugs have a risk of ototoxicity. If you have cancer, loss of balance may seem preferable to loss of life. The other class of highly ototoxic drugs are aminoglycosides, powerful antibiotics used to treat severe infections with gram-negative bacteria. These infections are resistant to most antibiotics, so it may make sense to use an aminoglycoside (such as gentamicin or streptomycin) to control potentially dangerous bacteria. If your doctor recommends an ototoxic medication, ask questions. Not everyone experiences permanent vestibular loss, and the risks may be reduced by taking only the lowest necessary dosage or by working with your doctor to limit the number of days you take the medication.

Nag, Nag, Nag

There's no cool way to say this, but please be careful. The lesson I took away from this research—indeed, the lesson that Dr. Strang spends his life teaching—is that we tend to worry far too little about the effects of alcohol and drugs on balance. That makes me worry for the people I know and love. Scratch that—it makes me worry for everyone. A stumble caused by dizziness could simply wrench an ankle, or it could result in brain injury after a fall down the stairs. I'm not saying to avoid alcohol or medications, but now that you know their effects on the center of your balance system, be aware. Ask for help if you feel dizziness or vertigo. Be careful.

9

Of Helicopters, 3-D, and Queasy Cam

Cybersickness

L ONG BEFORE SMARTPHONES AND iPADS, before 3-D televisions and laptop computers, even before Nintendo game systems,* there were flight simulators. These precursors of modern-day video entertainment really began to take off in the late 1970s, when digital technology started to become more realistic and responsive. While most of us know *Flight Simulator* as a cool game, real simulators

---

*But not totally before Nintendo. The Japanese company was founded in the late 1880s, and they originally produced *hanafuda*, or "flower cards." These cards—used for several games, including Koi-Koi, which is highly popular in Hawaii—don't look like typical playing cards with suits, numbers, and face cards. Instead they are unbendably thick and tiny—only about two inches tall—and depict flower groups that correspond to the twelve months of the year. While it takes some time to get accustomed to the physically smaller and thicker size of the cards, the illustrations are beautiful and the basic games are quick and fun. I found modern-day Nintendo hanafuda cards online at Amazon. (I also found a more card-like "extra large" Hawaiian version, which was much easier for my clumsy hands to manage, at HanafudaHawaii.com.) It's gratifying to know that the creators of Mario and Luigi had such a classic beginning.

began as tools for training pilots and were used extensively by the US military. Even now, simulators allow pilots to test their ability to fly straight in hazardous scenarios—including foul weather, combat, or instrument failure—without the risk of death if they fail. But simulators carry serious risk of a motion sickness–like reaction dubbed simulator sickness.

I imagine that simulator sickness was a nasty surprise to the early users, many of whom were seasoned pilots with thousands of hours of flight experience. The problem is the same one that still plagues pilots, as well as users of virtual reality and people who watch 3-D movies: sensory conflict. While the visuals in the flight simulator mimicked what pilots saw in a cockpit, the bodily sensations the pilots experienced when flying in a real jet were absent—no gravity-pull of acceleration, no proprioceptive cues marking a turn, and no changes in the physical perception of up, even when simulating a roll. And remember, there were no home computers back then, no video games. Those early simulator pilots were virtual virgins. The only screens in their lives were the televisions in their living rooms or the silver screen at the movie theater. They weren't prepared for visually induced motion sickness. And they got very, very ill. Researchers noted symptom after symptom—dizziness, vertigo, confusion, disorientation, headache, cold sweats, and some nausea—and recorded anecdotes, including this note: "On one occasion an instructor had to get out of his car on the way home and walk around in order to regain his equilibrium. Other instructors became conditioned to the simulator to the extent that the very sight of it made them sick."*

---

*From the 1960 article by Miller and Goodson titled "Motion Sickness in a Helicopter Simulator."

Interestingly, in the early studies, pilots who experienced the most simulator sickness were those with the most flying experience. When researchers looked at the rate of simulator sickness and experience, about 20 percent of the group of fixed-wing pilots who had logged fewer than fifteen hundred hours had symptoms, versus about 50 percent of the pilots who had logged fifteen hundred to four thousand hours. On the surface, those statistics make no sense, but there have been a few theories as to why this happened. Perhaps the more experienced pilots were older, so the real factor affecting simulator sickness might be age. Or perhaps the more experienced pilots had greater physical expectations from flying—their bodies had a stronger "sense memory" of what was supposed to happen in the plane, and when it didn't happen their sensory conflict was greater and led to more pronounced vestibular symptoms.*

Simulator sickness was (and is) most frustrating for pilots because symptoms could last for several hours after the simulation ended, which meant the rest of the day was shot in terms of their usual productivity and activity. However, nearly all pilots were fully recovered after a single night of sleep.

Helicopter pilots took the worst hit. Their rate of simulator sickness reached 77 percent, including flight instructors. It is believed that helicopter simulators create more illness than fixed-wing simulators because helicopters fly closer to the ground, leading to more visual flow—more virtual objects speeding by—and therefore to a greater possibility of visually induced motion sickness symptoms (like those created by the Vominator†).

---

*It was also suggested that the less experienced pilots might not want to admit symptoms for fear it would affect their flight status, but that reason was dismissed because it couldn't account for the overall vast number of sick pilots.

†I talk about my experience in the Vominator in chapter 1.

Researchers tried to capture and quantify the symptoms of simulator sickness, but there was no tool that accurately defined the scope of the problem. They made do with motion sickness surveys, but there are some important differences between motion sickness and simulator sickness. Compared with motion sickness, simulator sickness causes fewer gastrointestinal symptoms but more visual problems, vertigo, and disorientation. For example, about 75 percent of people who are seasick will vomit, but less than 1 percent of people with simulator sickness actually hurl. Another major difference is the effect of closing your eyes—symptoms lessen or disappear if you have simulator sickness but don't go away with seasickness. That's because simulator sickness is caused by involvement with a visual display. Take away that visual input and the cause of the illness goes away.

Despite those symptom differences, scientists were stuck using motion sickness questionnaires in their research until 1993, when Dr. Robert S. Kennedy and his colleagues published a study describing a new method for quantifying symptoms of this new technology. They called it the Simulator Sickness Questionnaire (SSQ). The SSQ is considered the gold standard for evaluating symptoms in any situation involving visually induced motion sickness. Distilled from data from 3,691 "hops" (individual flights in a simulator), the SSQ is a quick list of sixteen symptoms: general discomfort, fatigue, headache, eye strain, difficulty focusing, salivation increasing, sweating, nausea, difficulty concentrating, "fullness of the head," blurred vision, dizziness with eyes open, dizziness with eyes closed, vertigo, stomach awareness, and burping. Participants in research studies are asked to rate how much each symptom affects them "right now" (at the time of the study), from "none" to "severe."

The SSQ has proved to be a sturdy and reliable questionnaire, and it is still used today. Back when Dr. Kennedy developed the questionnaire, it would have been difficult to imagine all the permutations of simulator sickness there would be by 2017. People have reported feeling motion sick from the dimensionality on their smartphone screens, from fast-cut or shaky-cam movies, from 3-D movies, and from all forms of virtual reality. For each type of device, we have a different name for the symptoms, including phone sickness, movie sickness, and virtual reality sickness. The one thing they all have in common is the use of visual digital content displayed on a screen. In this way, they all can be grouped together and called cybersickness, or, as I have been thinking of them, screen sickness. So many different sources, all leading to similar symptoms. But it's interesting to examine them individually to try to understand why they affect our vestibular system.

Movie Hurl

By day, Keith Wiley works in data sciences; by night, he is a subtly subversive sufferer of cybersickness. Several years ago, Keith noticed that more and more movies were making him sick. And it's not just Keith. Moviegoers and reviewers across the United States are experiencing nausea and dizziness caused by a cinematic style that has come to be known as shaky cam or, more pejoratively, queasy cam or jitter cam.

Back in the days when movie cameras were both enormous and very heavy, moving one about was a chore, so they weren't moved all that much. That's why most scenes in older movies are stable. Further, action scenes were created by choreographers, actors, and stunt performers—one actor punched, another actor received the blow, both actors brawl. We saw it all continuously unfold.

Now, digital technology has improved, cameras have gotten much smaller and lighter, and movie action can be simulated instead of performed. There are lots of tricks directors and cinematographers use to get our attention and increase the velocity of scenes on the screen. The most obvious tactic first came to wide public attention in the late 1990s with two movies: *Saving Private Ryan* and *The Blair Witch Project*. These films were made with handheld cameras to create shaky, jagged, rough action that evoked emotions in the viewer. For example, in the opening battle scenes of *Saving Private Ryan*, we got the feeling of chaos, of the harsh and fierce nature of combat. The technique was used sparingly, like a sprinkling of Parmesan cheese on a plate of spaghetti, to enhance the story. In *The Blair Witch Project*, that shaky camera effect *was* the movie. The movie is about three people out to shoot a documentary about the fictional Blair Witch, and the story is told through the "recovered footage" from the fictional characters' cameras. The footage was all shot with handheld cameras, which means there's a lot of natural shaking, plus wild camera movements as the characters run and crash through the woods. This was the movie that put the shaky camera technique squarely in the faces of viewers. For the most part, we loved it. But a small number of viewers had a different reaction. A *Washington Post* article from the time said, "'The Blair Witch Project,' a new documentary-style horror flick filmed with a shaky hand-held camera, is making viewers stomach-quivering, skin-crawling, bathroom-dashing sick. Dizzy spells, queasiness, cold sweats and occasional vomiting have been part of the experience for some who've seen this sleeper hit."

But shaky camera is just one cinematic technique that can make us sick. The way moviemakers shoot and cut action scenes can create a sense of energy, of fast-paced wildness, or convey that things are out of control. For example, many action sequences today are not shot perfectly in frame, use various camera angles and setups to capture the action, and are edited with lots of quick cuts. This causes a viewer to lose the geography of a scene, making it difficult to keep track of exactly who is doing what to whom and what the outcome of the action is. The effect is to evoke a strong feeling of excitement and a sense of violence, without clearly showing the action. Among the worst cinema offenders is *The Bourne Ultimatum*, which leads many lists of most nauseating movies. It has scenes so short, choppy, and sick-making that it had some people vomiting in theaters.

If the shaky-cam technique enhanced the goals of both storytelling and emotionality, then it might be worth sitting through if you are prone to motion sickness (provided you dose yourself with Dramamine beforehand). But film historian David Bordwell famously called out *The Bourne Ultimatum* in a 2007 blog, saying, "In this case the style achieves a visceral impact, but at the cost of coherence and spatial orientation." Famed movie reviewer Roger Ebert, in an essay entitled "The Shaky-Queasy-Ultimatum," similarly wrote: "What is crucial (the "vomiting point," we could call it) is apparently when a film doesn't vary its pace, but is largely made of short hand-held shots, edited together by quick cuts that ignore spatial continuity."* You may not have noticed these crazy cut techniques while viewing movies before, but after watching this documentary, you won't be able to miss them.

---

*If you want to see what they mean, and you don't want to watch the whole movie again, you can see some examples on YouTube with *Unsteady—A Short Documentary on Shaky Cam in Action Scenes. Unsteady* can be seen here: www.youtube.com/watch?v=ebRIyyzbVkA.

Shaky-cam and quick-cut movies have the potential to make some people sick for the same reason that an optokinetic drum* makes people sick, the very reason we become motion sick at all—a sensory conflict between what our eyes are seeing and what our bodies are feeling, as translated in the brain. In the case of watching shaky-cam movies, our eyes tell us that we are moving, but our bodies are most definitely sitting in a cushy seat *not* moving. This sensory conflict can initiate nausea and dizziness.[†] Watching movies is an immersive experience; they take you out of yourself for a while and make you *feel*. Action—the wild scenes and unstable images—can be so realistic that your brain is fooled into thinking that *you* are experiencing the motion. That's especially true for movies shown on a large theater screen. Motion sickness experts say that the bigger the screen, the worse the movie sickness. Big screens flood your field of view and create a more immersive cinema experience that can send stronger sensory signals. That's why movie sickness is worse when you sit closer to the screen or if you see a really, really big movie, such as the ones shown at IMAX theaters. Conversely, the risk is lower when you watch on an iPad or TV screen in your home. Those small screens are less immersive and the visuals less believable, especially when you can easily see the edges of your iPad or the wall color surrounding your TV.

As with seasickness, carsickness, and other forms of motion sickness, some people are more susceptible to movie sickness than others. Data scientist Keith Wiley is one of those people, and he's not taking the proliferation of shaky-cam movies lying down. In 2008 he created the website Movie Hurl, which helps fellow movie

---

* Again, I reference the infamous Vominator of chapter 1.
†Of course, movie sickness is the inverse of regular motion sickness. On a boat or in a car, your body senses movement while the eyes see the stability of the vehicle. With shaky cam, the body senses stability while the eyes see immersive movement.

sickness sufferers predict which movies are most likely to make them feel sick. The website motto: "Learn which movies will make you motion-sick before you go to the theater and waste your money!"

On the basic level, Movie Hurl acts as a generic database of sickness ratings provided by other nausea-prone moviegoers. Anyone can rate a movie on a four-star "hurl factor," with one star meaning "no motion sickness," and four stars meaning "practically or actually vomited." The genius of this database is that it isn't just a reflection of Keith's own sickness reaction. Rather it collects ratings from anyone who wants to submit one. When I spoke with Keith, the database covered about 450 movies rated by more than a thousand users, with nearly ten thousand total ratings. Look up the title of any movie, and you'll find a hurl rating that is the average of all user submissions. Better yet, the hurl predictions can be personalized. Registered users* can rate their own movie sickness susceptibility—do you get really sick really often, or do you have a relatively strong stomach? Then, when you look up a title, the database returns personalized, more accurate hurl ratings.

But even the generic ratings are valuable for a quick view of which movies are likely to cause discomfort. For example, at the time I visited the site in 2017, the "Worst Movies Hall of Shame"— movies with the worst cybersickness ratings—included (in order):

1. *The Blair Witch Project*
2. *The Hunger Games*
3. *District 9*
4. *Beasts of the Southern Wild*
5. *Captain Phillips*
6. *Cloverfield*
7. *The Bourne Ultimatum*

---

*Registration is free!

Personally speaking, that list feels exactly right. I couldn't finish watching *The Blair Witch Project* because of queasiness, and I could deal with only the last two of the Bourne series, watching them on DVD in my living room. If I had known about Movie Hurl before I attempted to watch them, I could have prepared myself.*

But Movie Hurl isn't just a public service website. As Keith writes:

> This website is nothing short of an outright attack on the movie industry. If this website can drive enough viewers away from the theater, it is my hope that directors will stop making films this way. Alternatively, since such aspirations are bold to say the least, this website should at least permit people to save themselves horrific experiences and gobs of money.

As a sometime movie sickness sufferer, I can get behind that mission.

The Extra Dimension Makes a Big Difference

Shaky cam isn't the only thing that makes some movies more likely to cause illness. My sister can attest to that—she is highly susceptible to carsickness and airsickness, and she also seems to be sensitive to screen sickness. I once had to pull my car over to let her hurl after we enjoyed an IMAX 3-D showing of *Jurassic World*. On Movie Hurl, this movie rates only two stars ("mild motion sickness"). However, when we saw it, all the action and tooth-chomping gore was on a really, really big screen. The screens in dedicated IMAX theaters are wider and higher† than average

---

*The preparations would mostly be mental. If I am prepared for cybersickness, I'm not taken by surprise. Because I'll be monitoring my head and gut for discomfort, I can simply look away from the screen before the symptoms get too bad.

†The screens extend from floor to ceiling, but the actual height depends on the theater. In

theater screens, and they are curved for enhanced visual immersion. IMAX theaters also feature dual projection for extra clarity and reflective screens for heightened realism. Combine all those features and you have created the optimal viewing experience—and the perfect breeding ground for nausea, disorientation, and other motion sickness symptoms.

Early on, IMAX theaters were plastered with warnings about the possibility of feeling dizzy while watching an IMAX movie. I haven't seen those recently. Maybe I just haven't noticed them, but it wouldn't surprise me to learn that the warnings are no longer necessary for most people. Experts I spoke with say that as we all become used to seeing such large screens, we adapt to the sensations and eventually feel less and less sick. There are many fewer reports of illness in the theater now than when the experience was fresh and unexpected. Still, this format is more likely to create dizziness and nausea than a standard movie theater screen.

And my poor sister's nausea may also have been augmented by the 3-D aspect. Traditional movies are 2-D, meaning that no matter how high-quality the screen and projectors are, the image is still flat on a screen. While that can be highly immersive, it is nothing compared with 3-D movies. With the help of special lenses and stereoscopic projection, 3-D movies make images pop off the screen. This technology has been around for a long time—the first 3-D movie, *The Power of Love*, debuted in 1922, and the first color 3-D movie was the 1953 Vincent Price horror film *House of Wax*. But the technology has advanced so far that the dimensionality and

---

true IMAX theaters (versus regular theaters retrofitted for IMAX movies), the screen is at least fifty feet tall (about the height of a five-story building). But that's just the minimum height. There is one theater in Darling Harbour (Sydney, Australia) that has a screen nearly one hundred feet tall.

realism can be truly overwhelming for people who are sensitive to motion sickness.*

While sensitive people know the power of visual images, actual scientific research into the relationship between motion sickness and IMAX screens, 3-D movies, and other forms of visual stimulus proves the connection. For instance, researchers at the Department of Public Health and Infectious Diseases at Sapienza University of Rome (Italy) have been looking specifically into the health effects of 3-D movies. They set about to ask the question: Are there side effects to watching 3-D movies? The answer was a resounding yes. The process they used was simple. They asked about five hundred healthy adults to go to a theater and view either a 2-D movie or a 3-D movie of their choice.† (The choice was important because when one of your endpoint measurements is nausea, you don't want to force sensitive people to watch a horror or extreme action movie if that's not part of their usual viewing habits.)‡ Within three weeks, the subjects repeated the process, watching a film

---

*Nausea and dizziness aren't the only health reactions to the realism in 3-D movies. Physicians at the Ohio State University Wexner Medical Center treated a woman who experienced extreme heart palpitations, nausea, and vomiting while watching a 3-D action movie. It got so bad that she had to leave the theater. She went to the hospital and was diagnosed with takotsubo cardiomyopathy, also known as stress-induced cardiomyopathy or broken-heart syndrome. This is a temporary condition in which part of the heart enlarges and has a harder time pumping blood. The other part of the heart is either normal or pumps more strongly. Broken-heart syndrome is caused by an extreme emotional reaction of any type—good or bad. (Japanese physicians first recognized the disorder in 1990 after recognizing a pattern of disease that started after an earthquake.) What this tells us is that 3-D movies can elicit enough sense of reality and emotional connection that they can, literally, break someone's heart.

†The researchers noted which movies were chosen. The most popular 2-D movie watched by the research subjects was *Sherlock Holmes: A Game of Shadows*, and the most popular 3-D movie, chosen by nearly 80 percent of viewers, was *Puss in Boots*. (The study was conducted in 2011.)

‡Given that most subjects watched *Puss in Boots*, I think it is safe to say that any nausea or other symptoms of discomfort came from the 3-D movie effects and not from any extreme violence or gore in the movie.

of the other display type. By the end of the experiment, the participants had watched one 2-D movie and one 3-D movie. More than half reported some motion sickness–like side effects, including nausea, disorientation, and headache. While people with a history of motion sickness or headaches were more likely to experience symptoms, overall a participant's experience of symptom intensity was 8.8 times higher after viewing a 3-D movie than before the movie began. Watching a 2-D movie caused symptom intensity to only double. In other words, the risk of symptoms occurring was more than four times greater while watching a 3-D movie than it was watching a 2-D movie.

That made me wonder: If convincing, highly immersive movies can cause cybersickness, what happens with the ultimate in visual and psychological immersion—virtual reality? And can the power of VR ever be used for good? I dug a little deeper into the experience . . .

10

Beyond Gravity

Virtual Reality

I AM IN AN ITALIAN VILLAGE with the most annoying carousel music playing in a continuous, inescapable loop. The few people around are all ignoring me, but that's OK. I'm here not for the culture but for the view, which is mildly breathtaking. I want to walk through an archway toward the sea, but I keep ending up outside of a building that looks like a cathedral. It feels like I am moving around in someone else's dream—everything looks animated, and yet nothing looks familiar. Worse, I seem to have very little control over where I am going. I press forward, but I move sideways.

"I find that knob very difficult to use, I must say," says Dr. Laurence Harris, professor of psychology, kinesiology, and health sciences at York University in Toronto, Canada. This was the last stop on my tour of his multisensory perception labs. "Try wiggling the control around a bit. See if you can pull it up and down, rotate around, and push it forwards and backwards."

This is my first experience with virtual reality (VR)—the first of three I would test while researching this chapter—and I have a temperamental control stick. That doesn't stop me from loving every minute of it. I have always been a sucker for any kind of mind-machine interface. I remember my first experience with a video game that was more interactive than *Pong*. It was called *Zork*, a text-only adventure fiction game that I could run on my home computer. There were no graphics, just words on a screen that described a scene and the mission. The user (me) could type in commands, which the computer "understood" via sophisticated text-parsing software. It was fantasy literature that could be played like a one-person *Dungeons & Dragons* game. I tried Nintendo games a couple of times, and I found that I was prone to becoming obsessed with visually engaging games. For the sake of my marriage and my sanity, I decided it was safer for me not to own a gaming system. My biggest visual vice right now is going to see 3-D IMAX movies. I have never become sick or dizzy in an IMAX theater, and it never gets old. Needless to say, I love virtual reality Italy.

For this experience, I climbed up some metal steps to a large platform that held a large, curved screen; a bank of 3-D stereo-scopic projectors; and a chair with controllers. After I was seated, the chair moved forward so that it felt as though I were sitting inside the curve of the screen, with the sides winged out to make sure that all I saw was screen, even in my peripheral vision. This was full visual field virtual reality.

It is difficult to accurately describe the experience of virtual reality to someone who hasn't experienced it. It is a re-creation of the world. When you look at "real" reality, it is all around you. It is the ceiling and the floor,* and a 360-degree panorama around

---

*Or sky and ground, if you get outdoors more than I do.

you. Everything is three-dimensional, so you can discern distances between yourself and other objects, as well as the distances between one object and another. Virtual reality mimics real reality. It's all there—three-dimensional scenes that span your entire field of view. You can move around within the VR environment (in this case, with the help of the tricky controller), and the scene adapts, changes. For example, when I was in the virtual Italian village, I could "walk" between buildings and find myself on a patio over-looking the sea. If you have ever seen a 3-D movie, imagine not just watching the action on the screen but actually walking through the screen into the scene. You could be on Mount Everest (with-out the freezing cold and lack of oxygen), stand next to Captain America, or duck the kicks of a Kung Fu Panda. The technology isn't quite at the hyperreal level of walking into movies yet, but that can't be too far off.*

As with all VR experiences, including 3-D movies, I could only experience this slice of Italy by wearing a special set of goggles. VR goggles differ in appearance from one system to another. Most are enormous and covered my face from eyes to scalp. Others, such as the HoloLens headset, are more like thick halos with lenses.†
Without goggles, the projected VR images don't look like any-thing special—everything is distorted, blurry, doubled, or off-color. Lenses in the goggles take the projections and focus the images differently to each eye, creating a highly realistic stereoscopic 3-D effect. When the images are projected on a large, wide, wraparound screen like the one Dr. Harris has, or in the wraparound goggles

---

*Of course, the ultimate movie example of virtual reality mimicking real reality is *The Matrix*, in which people are unknowingly plugged in to a virtual reality world they grow up believing is the real world.
†The HoloLens goggles remind me of the appliance worn by blind engineer Geordi La Forge on *Star Trek: The Next Generation*.

themselves as they are in other systems, there are no confining edges, as there would be if you were merely watching a 3-D movie on TV. The VR scene fills the entire field of vision, making a very convincing version of reality. It tricks the brain into believing that what you see is where you are—you believe that the image is real, and you are *there*.

DiVE in, Both Feet

My problem with VR Italy is that I didn't end up feeling any motion sickness (as I hoped I would for the sake of this book); neither did I feel a sense of dizziness or disorientation, which is a documented possible symptom after a VR experience. So I went on the hunt for a more intense VR experience. After all, how could I possibly report on cybersickness if I don't know what it feels like?

My search took me to Duke University in Durham, North Carolina, one of the national hotbeds of all things science-y and technical.* After a few phone calls and e-mails, I was able to wrangle a private tour of some of Duke's VR research space.

That's how I end up standing inside a giant glowing cube, with surgical booties on my feet and large goggles over my eyes. I'm nervous, partly because I have no idea what to expect from this experience and partly because it already feels like the opening scene of a science fiction horror movie.† Then I hear the best sentence this journalist could hope for:

"I'm going to do my best to make you sick today."

---

*That's not hyperbole. Research Triangle Park (RTP), the largest research park in the country, is located within the geographic "triangle" of Raleigh, Durham, and Chapel Hill. It has been called the Silicon Valley of the South.
†If it were a movie, this would be the part where audiences would scream, "Don't go in the cube!" Because we all know bad things probably happen in the cube. But, as always, the woman goes in the cube.

My partner in the cube, and the man who made the welcome threat of nausea, is Dr. Regis Kopper, assistant research professor of mechanical engineering and materials science, and—most importantly for me—director of the Duke immersive Virtual Environment (DiVE). The cube—three meters by three meters by three meters (that's nearly ten feet all around)—is a room in which all sides are screens for the rear-projected stereoscopic 3-D projections. This is full-immersion virtual reality. It's better than just being close to a big screen. Here I didn't just walk into a 3-D movie; I am surrounded by it, a part of it.

I knew that much before my visit to see Dr. Kopper at Duke University, but I didn't completely grasp the enormity of the experience. I thought it would be similar to watching a 3-D movie projected on each wall all around me, a 360-degree movie. I was not prepared for this total immersion experience.

The booties over my shoes are necessary because the floor is a screen. Four walls, ceiling, floor—all screens.* Dr. Kopper closes the sliding glass door so that we are both enclosed in the cube.† He is younger than I expected from someone with *director* in his title. With his well-worn polo shirt, shock of dark brown hair, open expression, and shy smile, Dr. Kopper could be mistaken for an undergraduate. He has an intriguing accent that I can't quite place, not Spanish or French or German or Austrian.‡ He

---

*DiVE is one of only four room-sized cubes of this type in the United States. You might think that such a rare device would be highly protected and reserved only for scientists and engineers. But Duke recognizes that "facilities like the DiVE are 'watering holes' that bring people together." Anyone can participate in their weekly open house (first-come, first-served), or schedule a private tour by visiting http://virtualreality.duke.edu/visit/.

†At the controls outside the cube is David Zielinski, a research and development engineer at Duke University and a virtual reality expert. He is as important to the virtual reality illusion as the Great and Powerful Wizard was to the enchantment of Oz. He makes the illusion happen.

‡When I looked up his curriculum vitae later, I discovered that Dr. Kopper is in his early thirties. I also see that he got his undergraduate and master's degrees in Brazil, which explains his exotic accent.

comes across as a little reserved, but only until he starts talking about his work. Then he becomes as excited about the DiVE as I am. He hands me a controller wand and begins the software. Suddenly I am in a kitchen. An animated, cartoonish version, but most definitely a kitchen.

What is immediately remarkable about this experience is that I can walk around within the cube and the virtual reality scene remains true. There is an animated egg frying on the stove, a cartoon cat meowing as it walks on top of the cabinets, and even an animated cockroach crawling on the counter toward the sink. Every item can be picked up or moved, and the results follow the laws of real-world physics.* Using the controller wand, I can lift the virtual pan off the virtual stove, knock a virtual bowl off the virtual counter, open the virtual refrigerator . . . whoa! This is where the immersive 3-D experience pushes the amazement button in my brain. I pull the freezer door, and it opens out into the room so realistically that I take a step backward to keep the door from cracking into my head. That's when it really hits me: I am in an empty cube, but my brain has been fooled to the extent that I am ducking nonexistent doors. It is literally a spine-tingling moment.

The Wizard of DiVE† changes the scene so I am in a French cathedral, and I find myself moving in and around the architecture. As I walk among the topmost arches, I start to get a mild feeling of claustrophobia, which is a secondary source of amazement—I don't get claustrophobic in the real world, but I begin to experience

---

*Of course, one of the first things I try to do is kill the cockroach, but here—much as in the real world—cockroaches are indestructible. The cockroach is part of a study looking at levels of stress under different conditions. It seems that cockroaches, even animated ones, are a good way to elicit anxiety.
†That's what I've taken to calling the VR engineer, David Zielinski. I'm sure he will hate the nickname, but that's how I see him.

that "crowded out" feeling in a VR environment, when in reality the cube is empty except for me and Dr. Kopper.

When I was four years old, my parents took me to Disneyland. That visit remains one of the most powerful memories I have. I remember bursting with excitement and happiness, as if my human body couldn't contain the flood of pure joy I felt. Everywhere I turned, I saw something new and wondrous and experienced that bursting-with-emotion feeling anew. That memory is relevant because that's exactly how being in virtual reality makes me feel—a potent, almost overwhelming, sense of exhilaration. I ask Dr. Kopper if he ever got used to those feelings of wonder and joy. Did virtual reality become ho-hum? He tells me that despite guiding lots of individuals and groups through the experience, he doesn't often put on the goggles.

"If I'm in it too much, I get sick," he says. "Mostly headaches."

Even the Wizard of DiVE agrees that headaches are a problem when he spends too much time in a virtual reality environment.

I don't get a headache, but Dr. Kopper keeps his promise—he does make me a little queasy. I was fine in the virtual kitchen and even through most of the French cathedral. But there is a moment when I walk under an arch and my head goes through a virtual stone column. I immediately feel sick—not ready-to-hurl sick, but generally ill. My brain was fine playing make-believe with my environment, but I triggered sensory conflict warnings and illness when my head *went through* the column. Heads aren't supposed to move through stone! Talk about unexpected.

But this isn't the end of my cyber-nausea tour at Duke University. Dr. Kopper also allows me to test out the VR headset called Oculus Rift. I have been hearing about Oculus Rift for years through my nerd channels, but I had assumed it was little more than a revved-up video game system. Nintendo on steroids. Well, yes, it is that, but so much more.

The large VR headset of Oculus Rift feeds the virtual reality environment directly to the goggles—no screen, no cube—with what is called a head-mounted display. It is not self-contained, however; it requires a powerful PC to run the software, a controller, and a separate positional tracker. I won't pretend to understand exactly how it all comes together, but I can tell you the results. In a word: stunning. I get to play with game software called *Lucky's Tale*, in which I control an animated fox (named Lucky, in case that wasn't obvious) as he walks, jumps, and kicks his way around a forest environment trying to collect treasure. What makes this VR system different is that the graphics are displayed in the goggles versus on a screen. The environment feels hyperreal, as though the images are beamed directly to my brain.

I have a blast playing with Oculus Rift—until my Lucky falls into a deep crevasse. At that moment the queasiness starts, just like when my head went through the stone archway in the DiVE cube—both were unexpected deviations from what I expect by interacting in the real world. The *real* real world. As Dr. Kopper explains, it was moving in a "wonky direction" that made me feel nauseated.

"It has a lot to do with the feedback and what you are expecting," Dr. Kopper tells me. "Your mind and body are coordinating the experience you are having, and then something unexpected happens that gets you off-balance, literally."

Describing the Indescribable

I have spoken to a lot of people who don't quite understand what virtual reality is and how it is different from, say, a really good high-resolution video game. But comparing VR to a video game is like comparing a TV show to a comic book. Yes, they both tell a visual story, but the experience is entirely different. Incomparable.

Until my first walk through the virtual Italian courtyard, I certainly didn't understand the concept of full-imagination immersion, that feeling of letting myself believe that other worlds are possible and that I'm temporarily living in that alternate reality. It's like a deep dream that you want to return to after waking. I realize I sound like I've been inducted into some sort of cult, but it's really not a creepy thing. Still, all my words seem to make VR sound a little wackadoodle.

Even experts who have spent their lives with virtual reality have a difficult time describing the experience to the uninitiated. Trust me, I asked. The most common response I got was, "You know who you should talk to? Jason Jerald. He knows everything about virtual reality."

So I did.

Dr. Jason Jerald has an impressive résumé, beginning with his PhD in computer science from the University of North Carolina at Chapel Hill. He is cofounder and principal consultant at NextGen Interactions, a company offering high-level virtual reality product and software development. He has worked with NASA, Oculus, Valve, Raytheon, Defense Advanced Research Projects Agency (DARPA), IEEE Virtual Reality, and many other organizations. Dr. Jerald is also adjunct professor at both Duke University and the Waterford Institute of Technologies in Ireland. Oh, and in his spare time he wrote one of the seminal books for designers and engineers called *The VR Book: Human-Centered Design for Virtual Reality*.

"Virtual reality is something you can't explain in words*—you have to experience it," Dr. Jerald told me. "I came up with the term 'visceral communication,' technology communicating to you at a visceral, or gut, level."

I recognize that I was privileged to have had all those VR experiences, first with Dr. Laurence Harris at York University in Toronto (the Italian piazza) and later with Dr. Regis Kopper at Duke University (the DiVE cube and Oculus Rift). So I asked Dr. Jerald what would be an example of a VR experience that most people could enjoy.

"Well," he said, "you've tried Google Cardboard, right?"

I virtually smacked my palm on my forehead and changed the subject as quickly as I could. Google Cardboard! I had heard about it in passing, but I hadn't looked into it. For reasons I can't explain, I thought it was either a scam or a case of overinflated technological promises, so I ignored it. That was my mistake. Google Cardboard is a virtual reality platform that uses a smartphone as the screen, which is viewed through a set of stereoscopic lenses set in a cardboard viewer that you strap onto your head. You start by downloading a virtual reality app onto your smartphone; then slide the phone into the viewer and watch for the magic.

To see how this home-style experience compares with the pro-level VR devices, I purchased a viewer from Amazon.† The most difficult part of the experience was choosing an app that would best demonstrate the VR quality of this tiny device. There are many to choose from now, but not all apps work well with every phone model. I tried a popular horror game and found that the stereoscopic screen didn't line up properly on my phone, so I always

---

*How nice to have my sense of verbal inadequacy confirmed by this expert!

†At the time I ordered in the summer of 2016, it was possible to get the basic Google Cardboard for less than ten dollars online.

saw two flat images side by side, instead of the one merged 3-D image.* When I found something I liked, something that promised stunning graphics, I loaded it up and tried it out.

While the experience isn't as intense and all-encompassing as the others I tried, Google Cardboard (and other smartphone VR viewers) is a great way to get the feel for virtual reality. It gives you a peephole into the virtual world, allowing you to have that "visceral communication" Dr. Jerald spoke of. I say "peephole" because, compared with the larger devices, the smartphone VR has a smaller field of view, so you don't get that "I'm really there" quality. But you can turn your head in any direction—360 degrees around, up, and down—and you will see the corresponding portion of the VR world. For example, if you are in a natural forest environment and look up, you'll see sky and flying birds; if you look down, you'll see dirt and leaves and roots.

That 360-degree view (called the field of regard) is unlike the way we have always been taught to view screens. We're more used to the static one-direction view, like TV, which delivers only a single point of view. For this reason, it takes a little while to adapt to the new VR way of seeing the world. I kept forgetting to turn or move my head, which defeats the purpose of the concept. With a little experience, it became second nature to look around. I can sit on my couch and examine an underwater coral reef.† Or take a ride on the newest roller coaster at Cedar Point.‡ The bottom

---

*Someone wrote instructions online about how to fix the problem, but I was looking for "amazing" right out of the box, not something I had to work for. (As I reread that last sentence, I realize how VR-spoiled I have become. Heaven forbid I should have to work through five steps to experience something that my grandmother would have called a miracle.)

†I found this on the Within app, a collection of cinema-quality VR experiences, including comedy stand-up routines, music videos, and a bunch of visually beautiful short films and documentaries.

‡The Cedar Point VR app, which takes you on the Valravn roller coaster, billed as the

line: If you want a taste of VR, give this smartphone tech a try. It delivers a basic VR experience.*

Virtual Motion Sickness

Of course, the virtual reality experience is also likely to bring on balance problems—both visually induced motion sickness (VIMS) and actually falling over.

Virtual reality causes symptoms similar to other screen sickness, including motion sickness–like nausea and a good dose of vertigo, dizziness, disequilibrium, and disorientation. However, VR makes more people sick, with more extreme sickness, than other forms of screen sickness. Experts in the field believe that's because virtual reality is such an all-encompassing, immersive environment.

"In virtual reality," says Duke University's Dr. Regis Kopper, "one of the main differences compared with video gaming is that you are completely surrounded by the imagery, instead of just looking at the imagery at a distance. A big part of your view field is surrounded by moving images. There is a perceptual mismatch between what you are seeing and what your body is telling you. Your body is static, not moving, in a fixed position, but your eyes are telling you that you are moving . . . in the virtual world."

That immersion into a new world that surrounds us on all sides is the very thing that makes VR so magical. The brain is more

---

"tallest, fastest, and longest dive coaster in the world." Sit down for this one—it will make your stomach flip!

*There are lots of smartphone viewers available. Google Cardboard is just one brand and the most stripped-down version. Its main downside is that it really is made of cardboard, so it won't survive getting wet, and it is more easily crushed than plastic models. Some viewers come with a remote to make selections on a virtual screen. Others come with built-in earphones. There are even a few viewers that don't require a smartphone—they have a permanent screen built into the headset—but they are pricey and can run over $200. *Important note:* Make sure the viewer you choose is compatible with your phone. Because the phone needs to be placed inside the viewer, sizing is critical.

easily persuaded into believing that the virtual environment really exists when you can turn in a circle and experience being smack in the center of this new world. Good VR allows you to believe in the new reality and, by default, trust it the same way we have faith in the real-world ground beneath our feet. When that trust is broken in a virtual world, there is almost always a physiological reaction. In VR, if you fall down a pit, or if you walk through an imaginary wall (the way I did in the DiVE cube), chances are you'll feel something. Maybe not full-fledged nausea, but *something*. At the very least, your stomach will drop, the way it does on a roller coaster hill.*

"It has a lot to do with the feedback and what you are expecting," says Dr. Kopper. "So, you're coordinating the experience you are having, and then something unexpected happens that gets you off-balance, literally."

That's the joy of real-world roller coasters—that element of surprise that makes our bodies react physically. Our heart and conscious brain love the steep drops and tight turns, but our vestibular system is automatic. We can't ask it to calm down and enjoy the ride. It will create symptoms regardless of whether the thrill is real or virtual.

So it makes sense that people tend to get sicker when they are more engaged with the experience.

"When you let go of the grounding in the real world," Dr. Kopper told me, "it is possible that you are enjoying yourself so much in the virtual world that you are forgetting to keep track of keeping your body completely in check. You're more drawn into the virtual environment, and because of that, you get more sickness."

---

*Most people who ride on a virtual reality roller coaster experience physiological sensations similar to what they feel when riding a real-world coaster.

Of course, some people are just more prone to VR sickness. In his book, Dr. Jason Jerald lists the known factors that contribute to feelings of sickness in a virtual environment. VR sickness is more common if:

- you have a prior history of motion sickness.
- you are female. (No one has a good reason why.)
- you are an adult; the older you are, the more likely it is that you'll get sick.
- you are sick already (such as with a stomach virus).
- you have a history of migraine.
- you have limited experience with VR.

Another sickness factor is simply thinking about becoming sick. Typically, researchers measure how sick people become during a VR experience by giving them the Simulator Sickness Questionnaire (SSQ) before a VR experience, and then again afterward. The SSQ asks about various symptoms that may occur and requires the test-taker to rate how strongly they feel that particular symptom at the moment. In research, the difference between the "before" SSQ scores and "after" SSQ scores is thought to be due to the effect of VR. So if a person says he or she has no nausea at the beginning of the study but feels severe nausea after a VR experience, then the nausea is probably caused by virtual reality. Seems logical, right? One study, however, looked at the effect of the questionnaire itself. Half of the research participants did the "before" and "after" SSQs, but the other half took only the "after" test. The results showed that people who filled out the SSQ both before and after VR were, on average, sicker than people who only filled out the "after" SSQ. That suggests that simply *thinking* about getting sick is a risk factor for actually getting sick during a VR experience.

Caging the Adverse

Even immersion and that long list of risk factors aren't enough to define why virtual reality goes a step beyond other screen sicknesses. Those are simply the human factors. With VR, you must also factor in an enormous number of what Dr. Jerald calls the "system factors"—shortcomings in technology* that are capable of causing sickness in a user. For example, something called latency is thought to be the greatest cause of VR sickness because it causes extreme sensory conflict. Because VR is a 360-degree experience, people can turn their heads in any direction and expect to see a new scene. This requires a remarkable amount of planning and coding on the part of the software engineers. When latency is off, the visual cues presented when you turn your head lag or drag behind what you would expect to see. So if you are looking at a lake straight ahead, you would expect to turn and be able see the left bank of the lake; but problems with latency could make the scene move strangely, or create a kind of "swimming" sensation as the scene doesn't react the way you expect it to. You know that feeling you have in a dream when you are trying to run but your legs feel like they are stuck in mud? You might think of that as a latency problem in your dream.

Other system issues that can cause sensory conflict and sickness are tracking accuracy, precision, and position. In all, Dr. Jerald's book lists twenty different system factors that can create scenarios that make people seriously dizzy, nauseous, and off balance.

And if dizzy, nauseous, and off balance aren't already an abundance of possible health risks, Dr. Jerald told me about the one

---

*This refers to *current* tech shortcomings. Designers can't change human responses, but they are actively working to tweak or improve the way systems work to minimize human sickness. Chances are that by the time you read this chapter, many of the kinks in today's VR will have been worked out.

problem that no one is talking about: falling. Not only is the technology likely to cause symptoms, but also we interact with the virtual world in a way that can make us unbalanced in the real world. We can look up, down, and even spin. Remember being a kid and rolling down a hill or twirling on the grass? Remember how it used to make the world seem cockeyed for a bit and how that feeling could sometimes make you fall down? The same type of thing happens with virtual reality. Worse, with head-mounted displays, we become blind to the real world; all our visual input is from the virtual world, so we can't see when we might be walking into a chair or other obstacle. There's a pretty big risk of becoming so disoriented and off balance that you take a tumble and hit your head, break your arm, or smash your nose.

"So many demos [virtual reality demonstrations], by default, are standing experiences. Why?" Dr. Jerald asks. "Standing and walking experiences can work quite well if the system and application is designed for it, but standing should not be the default where there is not a reason for it. It seems like there's so much risk there—health risk, legal risk—if you put some crazy motion in there, put someone on a roller coaster . . . why do you have them standing? Why not put them in a seat? It would be so much safer." He pauses for a moment before continuing. "Maybe it's not an issue because we haven't seen anyone fall over and crack their head open yet, thank God, but it amazes me. Why not be safe and have people sit down before putting this crazy thing on their head?"

Why not, indeed. And he's right. In May 2016 electronics superstore Best Buy became the first and only US retailer where customers could try out Oculus Rift headsets in a limited number of stores. Yes, this intense VR experience is given to people who are standing up. I watched several YouTube videos of people having their Best Buy demos. While most people stayed relatively stable (and looked straight ahead, as if they were staring at a TV screen), some were noticeably wobbly. And some of the demonstrators didn't seem too concerned with the possibility of a user falling over. In fact in one video the tester was in a virtual world that appeared to be a glass elevator in a very tall building, with a massively, acrophobically steep drop-off. The demonstrator actually told the user, "Feel free to lean over!" The people standing around watching the demonstration laughed, but if the guy had bent forward and lost his balance, he could easily have cracked his skull on the sturdy demo table in front of him.

Harnessing VR Goodness

If virtual reality were just a gaming or entertainment platform, it would be easy enough to dismiss the technology as just another passing fancy. But virtual reality is expanding quickly, not only for fun but also for serious benefit.

"The velocity of change is insane," says Dr. Jerald. "Technologies I thought were five or ten years out are now one year out. The change is happening superfast. Right now it's all about entertainment, but I see other applications come up—medical, professional, training applications. I think those markets are going to dwarf the entertainment and gaming applications. Maybe not this year, but eventually."

I experienced one example of how VR might be used in the DiVE cube with Dr. Kopper. He is working with Olympic skeet shooters to see if VR could be used to train these elite athletes, what's called high-performance training. In the cube, a simulated clay disk is launched with variable speed and trajectory. It is my job, as the shooter, to lead the object, then virtually shoot it out of the virtual sky with my controller. I was terrible at this task, but

Olympic skeet shooters find no challenge with the standard version I used. However, Dr. Kopper and his team degrade some of the visual cues so that the athletes need to work harder to maintain their level of excellence, virtually.

"If you are training in suboptimal conditions," says Dr. Kopper, "you are going to have to push your whole self—cognition and behavior—so you can achieve high performance. Then when you go back to normal conditions you may have an edge."

One of the suboptimal challenges Dr. Kopper is testing is intermittent vision, when the virtual screen is blacked out for a certain percentage of time. In this case, the athletes see the images for only 20 percent of the time, and for 80 percent of the time the screen is blank. Specifically, with every second, the athletes get two hundred milliseconds of seeing the skeet and eight hundred milliseconds of blackness. If they can handle only 20 percent visual input, their performances under full-vision conditions should be improved.

"It's like a stroboscopic view of the environment," says Dr. Kopper. "We expect that we'll be able to help the athletes track better, more efficiently. We think that even the smallest incremental improvement will give them the edge when competing."

This type of training also allows trainers to monitor tiny behaviors that are usually invisible with real world viewing techniques. For example, with computers collecting data in the VR program, researchers and coaches can know exactly where the athletes aimed, where they hit, how much time they took to initiate motion (in milliseconds), when they started tracking the object, when they decided to shoot—everything that can enrich our understanding of what makes an expert an expert. It is also possible to measure and examine brain activity recorded through an electroencephalograph (EEG), then compare the EEG information to the shooting performance. Imagine comparing the brains of experts and novices and being able to actually see brain improvements as a shooter moves from being expert to Olympic caliber. You can't get that type of information just from watching video of a shooting range.

A Better Balance Cure?

One of the major risk factors for VR sickness is age. Kids seem to adapt more quickly to VR than adults do. And the older you are, the more likely it is that you'll get sick or fall over.

No one really knows why that is.

"Maybe kids are so excited by this [virtual reality] . . . maybe they get just as sick, but they're just ignoring it," says Dr. Jerald. "With older people, maybe it's not that they get sick more but that they don't trust the technology or are scared of the technology."

That could very well be true, especially in light of the study that showed that just thinking about getting sick could make research subjects feel sicker in a VR environment. And Dr. Jerald's thinking makes sense when you look at new research that uses virtual reality to help people maintain or regain balance skills. Perhaps when VR is presented as a health prescription, older people don't seem to get the same level of sickness they experience from VR gaming. Maybe the idea that virtual reality is a helpful tool transcends some of the preconceived notions of sickness.

Scientists around the world are experimenting with using virtual reality scenarios to help people regain balance after stroke, disability, illness, and the cruel infirmities of aging. But first let's look at the science behind standing still and moving—without falling down: kinesiology.

11

Think Not, Do

Psychology in Kinesiology

ONCE THE THREE MAJOR BALANCE inputs are in place,* we still need a way to put that information to practical use. That's where kinesiology comes into play. Kinesiology is the study of all factors relating to human movement, including physiology (how all the different parts of the body work and relate to each other to create a functioning person), psychology (how our emotions and thoughts—conscious or subconscious—affect the way our bodies work), and biomechanics (understanding how the laws of physics and mechanics apply to the human body).† Every action of your body—even the relative inaction of lying down—is the result of your physical self working within the guidelines of physics and possible body motions, and of your psychological self initiating and guiding the activity. If you are interested in balance,

_____

*Just to recap, they are the vestibular system in the inner ear, vision, and proprioception.
†This is not to be confused with "applied kinesiology," a controversial alternative medicine practice.

kinesiology is the study of how we sit, stand, and walk without falling over.

While very few of us will ever visit a kinesiologist,* researchers in this field provide the knowledge used in some way by most other areas of medicine. For example, let's say you break your left foot. After three months in a cast, the muscles in your left leg will have atrophied from lack of use. Another problem is that during the time you were in a cast, you had to learn how to balance using only crutches and your one healthy foot. Your body adjusts to having use of only one leg by shifting your postural alignment to rebalance and compensate for the functional loss. When the cast comes off, you will need physical therapy to regain strength in your left leg, relearn how to walk without a limp, and relearn proper balance. That physical therapy process is based on information provided by kinesiologists. With their understanding of human physiology, physics, and body mechanics, they have mathematical models for everything related to movement and balance. The physics of human movement is their thing.†

But kinesiologists go beyond mere mechanics. They also study how the brain and mind affect movement. To me, brain and mind have different connotations. The brain is the part that operates my body, even while I sleep. It is the thing that regulates breathing and other autonomic functions; it stores memories; it reacts to alcohol; it creates neurochemicals that affect my mood and my ability to concentrate; and it correlates all the inputs from nerves, vision, and the vestibular system to allow me to maintain balance without conscious thought. The mind, on the other hand, is "me."

---

*You'll be most likely to see a kinesiologist if you are an athlete and need some mental or physical fine-tuning to reach optimum performance, or if you have sustained a serious injury and need to learn how to regain physical function.
†More about biomechanics in chapter 12.

The part that thinks, that makes decisions, that can get nervous at the thought of giving a speech and can overcome those nerves to walk out on a stage and speak anyway. Kinesiologists also look at those types of factors—the subconscious acts of the brain and the conscious thoughts of the mind—in terms of how they can affect body movement.

That's what makes kinesiology the final piece of the balance puzzle. Kinesiologists study how the body moves and how to make it move in a better, more perfect way, even when your own mind is fighting against you.

That Monkey Wrench That Is Your Brain

Professional golfer Chris Cox knows firsthand how thoroughly the mind can punk the body. He got the yips,* and they nearly derailed the budding career of this young athlete.

I had heard of the yips before, but what I had imagined was not at all close to reality. I know many casual golfers, weekenders who are happy to break 100.† They joke about having the yips as an excuse for a particularly bad shot, as in, "So it was an easy chip shot, but I got the yips and the ball ended up in the trees." They

---

*The problem is called "the yips," always plural, always with *the* before *yips*. The verb is singular, so you can *yip* because you have *the yips*. A golfer with the yips can be described as *yippy*.

†For those of you who know nothing about golf, "breaking 100" means that the golfer took fewer than one hundred shots to complete eighteen holes of golf. Pro golfers shoot in the sixties or seventies. That's really good. If you shoot in the eighties, you are probably athletically gifted or you've been playing since you were ten or you're retired. (If you're my husband, shooting in the eighties means taking the foursome and their significant others out for a celebration dinner . . . with champagne.) If you shoot in the nineties, that's a respectable number for a typical weekend golfer. If you shoot over one hundred, friends will say something like, "At least you had fun." It's meant to be consoling, but it's the verbal equivalent of averting your eyes. (I'm not sure how true any of this is. I have never golfed. This is just what I have gleaned from listening in on golfers' conversations, which tend to last at least as long as it took them to play the round they are discussing.)

speak about it so casually that I never thought of it as a serious thing. And for my friends, any time the yips kicked in, the outcome was often so wild that I envisioned their arms and body making strange, exaggerated contortions that sent the ball far afield.*

None of what I had imagined matches reality. The yips are very serious for professional golfers. One multidisciplinary study found that among a group of more than a thousand professional golfers, more than one-third had experienced the yips. Every career golfer knows someone or has heard stories of pros who lost their game to the yips. The problem can take months or years to correct, interfering with a golfer's livelihood. But more than a few good golfers felt so frustrated and discouraged that they gave up the game entirely (not just professionally) after struggling with the yips. Furthermore, the yips aren't caused by some comedy skit of a crazy swing. Rather, the symptoms involve freezing and/or small jerks or tremors of the arms. But these microchanges indicate a block in the function of certain muscle groupings called synergies.

When we learn any new motor skill—whether it's golfing, dancing, or juggling—our muscles learn to work together in a specifically coordinated way. Early on, the activity seems difficult and our movements feel uncontrolled. An outside observer would describe our performance as stiff or amateurish. But with practice, two things happen: our brains create new neural pathways that control the activity, and we set up new muscle synergies—muscles that work together as a group, activated by a single signal from the brain. Each muscle can be part of many different synergies. So we have a synergy for walking, for skipping, for hopping, and for riding a bike. Those are four muscle synergies, each involving

---

*From what I've heard, the yips have been responsible for my husband sending balls into trees, sand traps (don't say "sand trap" to a golfer—they prefer "bunker"), a parking lot, and also smack dab into a gas grill on someone's back patio . . . twice. (Seriously. Twice.)

the same leg muscles. When we think *skip*, our brains activate the skipping synergy, and our bodies respond by skipping. You may have heard the term *muscle memory*. It's the same premise, but *muscle synergy* more clearly defines the role of the brain in activating those grouped, coordinated muscles.

I recently experienced a common, everyday example of the powerful, subconscious way muscle synergies guide actions. In my home bathroom, the hand towel hangs to the left of the sink. Whenever I wash my hands, clean my face, or brush my teeth, my left hand knows to reach out and grab the towel. *Water off, reach left. Water off, reach left.* It's a total muscle synergy honed by years of bathroom-sink practice. During a trip, I stayed in a house where the hand towel was folded on the counter to the right of the sink. For three days, every time I used the sink, my left arm reached out for the towel. It wouldn't be so bad except that the bathroom was quite small, and when my left arm shot out for the towel (which wasn't there), I smashed my hand into the wall. When I concentrated, I could reach to the right for the towel appropriately. It was only when I wasn't thinking specifically about towels that my brain's autopilot kicked in with muscle synergies. Another common example happens when you're driving a car, suddenly need to stop short, and find that you automatically reach out with your right arm to hold back a child in the passenger seat, even if there is no child there at the moment. Or when you reach without thinking to catch an item falling off the kitchen counter, only to discover that it was a very sharp knife.* You can probably think of dozens of similar activities, things you do, behaviors so ingrained that you find yourself performing them without conscious thought. Those are muscle synergies.

---

*Or is that just me?

The yips, in whatever form they take, are a psychological override of the muscle synergies. By the time a golfer reaches the level of a pro like Chris Cox, he or she has developed muscle synergies for each different type of swing, from the full-out arc of a drive to the pendulum sweep of a putt. Those synergies allow a golfer to swing the club and hit the ball the same way for years, leading to a predictable result. But in a game whose outcome depends on accuracy within a fraction of an inch, even tiny changes in a golfer's physical or mental performance can send a golf ball yards off course. The yips are a psychological monkey wrench thrown into the system. Stress, worry, distraction, or other intrusive thoughts block the activation of the appropriate muscle synergies; performance suffers; and the golf ball skitters off the mark. If that happens once, it's a fluke. If it happens over and over again, it's the yips.

Career Enders

According to Chris and his instructor and mentor, PGA teaching professional Greg Greksa, overcoming the yips is extremely difficult. I met them on a golf course in North Carolina on yet another sweltering August afternoon, one in a long string of days with a heat index over one hundred degrees Fahrenheit. The men had been out on the course for over an hour, but they were more than willing to take me out in a golf cart and play another few holes. I countered with a suggestion that we sit down in the air-conditioned clubhouse. That's where they told me about Chris's problem with the yips.

Chris is in his twenties, with a young guy's energy and enthusiasm. He has one of those All-American faces that will someday make him a popular choice for ad placement. He's a natural for a box of Wheaties. If they made a movie of Chris's life, he could convincingly be played by Chris Pratt; and his mentor, Greg, could be played by Russell Crowe with a stubbly beard. They love talking about golf, and their memories are amazing. Chris went into great detail about each shot he played for the final five holes in a tournament five years ago.*

It's tough to find golfers willing to talk about the yips, mainly because they don't want to invite the specter of the yips anywhere near their game. It's a jinx. But Chris was confident that his case of the yips was thoroughly cured, and even if they came back, he knew what to do to get rid of them again.

"For me," Chris told me, "the yips involved taking motion that should be subconscious, you shouldn't even think about it, and manually trying to manipulate it. Like putting. It's such a subconscious movement. It's so simple that most of the best players don't really think about doing it. For me, putting turned into something very manual, where I was looking at my hands and thinking, 'These are just things that are stuck to the club, that aren't doing the job.' But nothing I did worked. If I'm holding my putter and move my wrist a quarter inch, a small twist, that little change in mechanics could put the ball two feet off line. So now I move it again, trying to fix the original problem, then again, and again. Before I knew it, my putt was gone."

For Chris, trouble on the course turned into a belief that he couldn't putt anymore. And guess what? When he *thought* he couldn't, he *really* couldn't. For him, the yips morphed to create a mindset that became a self-fulfilling prophecy.

---

*From anyone else, the stories might seem tedious, but Chris's enthusiasm is infectious. I found myself rapt with attention, even though I didn't understand about half of what he said. I have a decent memory, but his was beyond. He knew which club he'd used for the second shot, how many yards the ball had flown, and how his opponent had done in those same holes.

While the yips can happen at any point in the game, they are most common when players are putting, in part because it is such a simple motion, but also because it is the gold shot, where the ball either goes in the hole (cue celebrations and visions of victory) or misses. If you mess up on a drive, you can be yards off-line and make it up with the next shot. Putting is literally hit or miss.

Chris's case of the yips developed when he was a star golfer in college looking to transfer to a different school for scholarship opportunities. It should have been in the bag, but he got yippy.

"All this pressure being put on my shoulders. I remember thinking to myself that 'I gotta make this happen . . . I gotta make this work.' But then it didn't work, and then it just snowballed to where nothing I tried worked. It became, 'I got nothing . . . this is getting out of control.' If you watched a video of me putting at the time, you wouldn't see a noticeable problem. But the things you can't see, what's going on in my head, the little tiny movements of my hand, those were really where the problem was."

It probably didn't help that another kid Chris golfed with in college also got the yips, but that guy didn't find his way out, and stopped playing. For Chris, it was panic time. That's when Greg stepped in and saved Chris's career.*

Why Are You Telling Me This?

You might be wondering about now what golf and the yips have to do with balance. While the yips are a recognized issue among golfers, the same problem can happen in any sport or physical activity. The yips have appeared in other sports that require fine motor control, such

---

*That's what Chris says. Greg is far too humble to make such a sweeping, honorific statement. But from the story I heard, Chris was a diligent student but probably would have been stuck without Greg.

as darts or pistol shooting. The condition has also been documented in between 15 and 60 percent* of performing musicians, especially pianists, although it is usually classified as "stage fright." When the symptoms affect overall performance on a larger scale (as opposed to fine motor coordination), the result is called—as every nine-year-old knows—"choking."† It can even happen with regular, everyday walking. In fact it probably has happened to you, but you didn't know why. Have you ever been walking down the street, knowing (or thinking) someone is watching you, and you think to yourself that you're just going to be cool and continue walking? If so, there's a good chance you didn't stay cool. Instead, you may have tripped or stumbled or stubbed your toe on a nonexistent crack in the sidewalk.‡

That's one direct way that muscle synergies affect balance, when something interferes with your basic activity wiring, causing you to stumble and maybe fall. Or when your synergies are wrong for a situation. Take stairs, for example. Over the course of our lives, we develop muscle synergies for climbing stairs that are very specific to stair height. In the United States, building codes say that stairs should have a rise (the height between steps) no higher than 7¾ inches. But every now and then, you'll find stairs that are ever-so-slightly higher than code allows, and you'll discover it when you stub your toes as you climb. You can see the step, but your synergies are so ingrained to 7¾ inches that your foot placement will be off. In extreme circumstances, this could lead to a serious fall. In

---

*Different research studies find different numbers. This range is so large because the studies were done in different countries, with different types of musicians, with different ways of asking about the yips. Even at 15 percent, that's a lot of musicians unable to play at their full potential.

†I survived grammar school sports by going along with taunts that I had choked during a game. *Choking* suggests that the performance problem is temporary. I welcomed being told that I choked. It was preferable to admitting that I was just plain bad.

‡The appropriate reaction in those times is to turn and look at the pavement behind you, frowning, so anyone watching knows that you stumbled on *something*.

the introduction to this book, I wrote about my mother, who fell backward going up three steps from the garage to the house. She was very seriously injured, and we are lucky she is with us still. We don't know what caused her fall, but it could have been this exact problem. The steps were in her new house—this was her very first time using them. The bottom step was a little higher than usual, which means that her synergies might have been out of sync for the stairs. And she was carrying a bulky item in each hand—bags of new towels she had just purchased for the move—which could have changed her center of gravity. It's possible that the combination of weight imbalance and unusual stair riser height was too much for her to compensate for, and she fell backward.

That explains *what* happens, but *why*? Why do we trip on that imaginary sidewalk crack when we think someone is watching? Why would a professional golfer suddenly be unable to sink a four-foot putt?

It Really Is Mind over Body

If there is anyone who knows about the intricate link between mind and body, it is Nicole Detling, PhD, CC-AASP.* She is owner and senior consultant with HeadStrong Consulting, which provides mental skills training for performance enhancement, and assistant professor in the Department of Kinesiology at the University of Utah in Salt Lake City. Dr. Detling has worked with famous performance artists and Olympic athletes. I don't use the word *dynamo* often, but it seems to fit Dr. Detling—I have spoken with her several times, but never when she was sitting in an office. She is always on the move. I asked her what is going on in our brains to make us yip, choke, trip, or fall.

---

*CC-AASP stands for Certified Consultant of the Association for Applied Sport Psychology.

"There are a few theories out there," says Dr. Detling, "but the one I believe the most is that we become very self-conscious when we feel judged in some way. Then, all of a sudden, the small, simple tasks that we do mindlessly, including walking, we have to think how to do. There's a phrase in athletics that goes, 'The more you think, the slower your feet.' When we become too analytical about it, when we think too much about the task, your body can't respond to all those details at one time."

Imagine having to verbally describe the process of walking. Breaking down that one activity to its component parts is daunting: You put one foot in front of the other, but put your heel down first, at the same time you're swinging your opposite arm as you roll from that forward heel to your toe. When that lead foot is flat on the ground, shift your weight from the back leg to the forward leg ... When you start thinking too much about any simple task, that movement becomes much more difficult to do.

Research from scientists in Japan and Great Britain supports the idea that having an audience or feeling evaluated causes changes in your physical performance. Some fortunate people thrive with an audience and find that their talents are enhanced when they are being watched. But others choke. Studies of skilled pianists showed that being evaluated by an audience caused deterioration in their ability to play, affecting both the technical and the artistic qualities of the performance.

Another study, published in 2015 in the journal *Nature*, reported on research into how this kind of social monitoring might cause performance problems. Participants were given a simple grip test that required them to modulate how tightly they squeezed their hand into a fist. Some participants were led to believe no one was watching them, and others believed there were two scientists watching and evaluating their performance. (In reality, none of the participants were being watched.) The results showed that the group that thought it had an audience exhibited more anxiety and squeezed harder than instructed. In fact, the more anxiety, the harder the grip, even when the participants were asked to use a light grip. While subjects were doing the grip test, they also were undergoing functional magnetic resonance imaging (fMRI). This test shows active brain use, namely which areas of the brain were used and which areas were shut off while the participants did the grip test. Interestingly, having an audience changed the way brains responded during the experiment. The most telling result involved an area of the brain called the inferior parietal cortex (IPC). This part of the brain is involved in the integration of kinesthetic information—being able to understand and use sensory information to perform specific movements. The IPC was *deactivated* when there was an audience—or rather, when the subjects thought there was an audience. This is huge! It basically means that when we feel watched, when we have an audience we think is judging us, a part of the brain that we need to perform our physical skills *is turned off*. In the study, that's why subjects were more likely to grip too hard if they believed there was an audience. In daily life, that's why an audience might cause a pianist to lose subtlety when playing or cause a basketball player to crash one off the backboard rather than having the ball swish gently through the net or cause a golfer to tighten his grip on a club and ruin his putt. Metaphorically speaking, it causes the sidewalk to crack just enough for you to trip.

Not a Multitasker

Are you a good multitasker? I like to think I am, but it turns out that multitasking only really happens at the macro level. According to Dr. Detling, at any one moment in time, the mind can only have one thought.

"We switch back and forth between thoughts very quickly," says Dr. Detling, "but if I take a snapshot of your brain, there is only one thought there at a time."

That creates a conundrum. If I'm an athlete and I can have only one thought, that suggests that I should focus on my performance; but if I focus too hard, I freeze up or get the yips. So what should my one thought be? Dr. Detling has an answer, broken down into three parts:

1. Develop an individual preperformance routine.
2. Free your mind for muscle memory/muscle synergy.
3. Redefine success.

Preperformance Routine

A preperformance routine is the activity that you do immediately before you are called to perform. This isn't wearing the same pair of socks every game for a season or making sure you always eat pasta the night before a race. Those come under the category of personal preparation or superstition. The preperformance routine lasts just a few seconds, and it comes right before you swing the golf club or shoot a basketball or put your fingers on the piano keys to play a concerto. It is a way of focusing attention on the activities to come but without obsessing on the outcome. It can help you forget about the audience for those critical few seconds.

It works because when you're thinking about your preparation—setting the club or adjusting the position of your feet—you're not thinking about possibly missing the basket. You're not thinking *Oh my gosh, I have to make this shot or I'll lose my scholarship, which means I'll have to leave college. My mom will be crushed. I have to make this shot.* Thoughts like those are a good way to ensure that you'll fail. With a preperformance routine, you deliberately occupy the freaking-out space in your mind with a specific set of activities.

I was fascinated by this information, but it turns out that all good athletes know about the importance of a routine. Watch for it the next time you see a game on TV. You'll see players with very individual sets of activities they do to prepare for the game/match/shot/show.*

"A pre-shot routine is like a fingerprint," golf pro Greg Greksa says. "Nobody's going to do that pre-shot routine the same. As a teacher, I have to let [my students] develop their own way of preparing to take a shot, and then it's my job to pay attention to the details of that routine. Let's say a golfer has a routine that takes seventeen seconds before the putter is in motion. If all of a sudden that pre-shot routine takes twenty-four seconds instead of seventeen, then I know he's gotten something out of communication."

Player Chris Cox compares a pre-shot routine to a signature.

"No one's signature is the same," Chris says. "If I try to copy your signature, it's going to look really bad because I don't have those same hand and finger movements. I haven't practiced or trained my brain to do it. But my own signature is fluid. It's subconscious."

---

*This concept was perfectly illustrated in that classic episode of *The Honeymooners* called "The $99,000 Answer," in which Ralph Kramden prepares to compete on a game show, answering questions in the category of Popular Songs. He enlists neighbor Ed Norton to help him prepare by playing songs on a piano for Ralph to guess. Ed has a preperformance routine that consists of arm stretching, knuckle cracking, and—most famously—playing a few bars of the song "The Swanee River." You can see the whole episode on YouTube: www.youtube.com/watch?v=KF6ITeJ7EM0. (Go to the 16:12 mark to get to the preperformance routine.)

Free Your Mind for Muscle Memory/Muscle Synergies

So the routine takes care of the seventeen seconds or so before the actual heart of the performance. Then what?

If athletes or musicians think too much about their actual performance activity *during* the performance, they can easily get caught up in details. You have muscle memory for performance; you don't need to focus on mechanics. And if you're thinking about the audience watching you, you'll disconnect that kinesthetic part of your brain. The key is to switch thoughts. Dr. Detling recommends that instead of thinking about what your body is *doing*, think about how you want the action to *feel*. For example, if you're a golfer, instead of thinking about the angle of your arms as you raise the club, think about feeling "smooth" or "fluid."

"Some people like to use what I call deliberate distractions," says Dr. Detling. "This means taking your mind somewhere else completely so that your body can do what it's been trained to do. For example, counting, singing the ABCs, singing the lyrics to your favorite song—ideally one in which the melody matches the ideal rhythm of your swing."

Think of it as having a mantra that keys in your muscle synergies. Let your years of practice and development speak for themselves.

Redefine Success

Before you even get on the course or court, before the concert or presentation, it's important to take the pressure off by redefining what success means. That was a critical piece of how Chris got over the yips.

"One of the things we worked really hard on is not judging a good putt by 'you sunk the ball,'" says golf pro Greg. "Instead, judge a good putt by, did you go through your process? Did you put a good stroke on it? If so, then you did your job. Once you hit a putt, anything can happen. The ball can hit a bump or you can misread the green or any number of things. Just because the ball doesn't go in, that doesn't mean you didn't do your job."

The key for Chris was changing his results-driven mindset into a process-oriented mindset.

"It's you creating your own expectations," he says. "That's what we did. It's a blank slate, what's good and what's not, and nobody can tell you you're wrong."

At first glance that might seem like a recipe for failure. Of course a golfer should want to get the ball in the hole, right? Well, yeah. But Chris was at the point where the yips affected even his practice sessions. It was almost like he had internalized a judgey audience, so every time he picked up the putter, his brain fritzed, his muscles tightened and spasmed, and his performance suffered. By redefining success, he got that nasty, judging attitude out of his head and found his game balance again.

Bringing It Back to Balance

The goal of this chapter was to highlight how the workings of your mind are intricately woven throughout every movement you make. For athletes, the struggle is always to keep their mind in the right place, with the right goals, so they can perform. For most of us, that's not an immediate problem. Most people don't have to think about walking; we have strong muscle synergies for the activities of daily living. However, for those of us living a "regular life," the psychology and mechanics of kinesiology will eventually become important. As we get older, all our balance systems slowly fail. If you experience a stroke or have Parkinson's disease or another movement disorder, then balance is an immediate concern. Kinesi-ologists use these lessons in athletic psychology to help create tests and devices that will help us stay on our feet, balanced, throughout our lives.

What kinds of devices are on the horizon? What would you think about wearing a piece of Iron Man's suit? Some scientists are already working on that.

12

Building a Better Gait

Mechanics in Kinesiology

TODAY I MET THE GUY who is creating a real-life Iron Man. Well, not *the* guy—one guy who leads a team of researchers in one of many labs worldwide devoted to creating workable exoskeletons. That's what Iron Man is, after all. When comic book hero Tony Stark puts on his enhanced suit of armor, he transforms into a better-than-human superhero. The suit is the embodiment of physical perfection capable of great strength and endurance, with genius Stark providing the brains to direct all that power.

In nature, an exoskeleton refers to an animal's hard external covering or shell, which provides structure and protection for the organs within. Human structure is formed by the bones of our bodies, but they are on the inside, leaving our more tender bits on the outside. Animals with an exoskeleton wear their "bones" (or, rather, bone equivalents) on the outside. Some examples are crabs, lobsters, and grasshoppers.* The Iron Man

---

*You might ask, *What about armadillos?* Interestingly they do not have exoskeletons, although it is a frequent misconception that they do. Armadillos have regular skeletons

suit is an exoskeleton, providing near-invincible protection and magnified physical brawn.

That's the inspiration behind the research of Dr. Gregory S. Sawicki, associate professor and associate director of the Rehabilitation Engineering Core, in the Joint Department of Biomedical Engineering at North Carolina State University in Raleigh and the University of North Carolina at Chapel Hill.

"We definitely are inspired by Iron Man–type, Hollywood-style dreaming," says Dr. Sawicki, "but we acknowledge that we're pretty far away from getting to anything like Iron Man."

Sure. But even Tony Stark had to start with a dream and a spring.

Robot . . . Derp!

Human beings aren't perfect.* Think about how many times you hear a comment along the lines of, "Of course I messed up—I'm not a robot!" When it comes to balance on two legs, however, humans beat robots every time.†

Take, for example, the 2015 DARPA‡ Robotics Challenge (DRC). The challenge began as a way to deliver humanitarian aid and supplies after disasters—nuclear plant leaks, earthquakes, terrorist bombings—when the environment is too dangerous or unpredictable for human intervention. After preliminary meets,

---

(inside their skin) and hardened plates made of osteoderm, or bony skin. If the bony skin totally surrounded the armadillo's body, it might be considered a second skeleton. As it exists, it is considered armor.

*Except, of course, for Hoda Kotb.

†Robots balance better if they have a low center of gravity or more than two legs. For example, an eight-legged spider robot (www.robugtix.com/spider-robots) or a four-legged super robot (www.bostondynamics.com/robot_bigdog.html) can have better balance than a two-legged human.

‡DARPA stands for Defense Advanced Research Projects Agency.

twenty-three teams from around the world participated in the robotics finals in 2015, where the robots would need to demonstrate their ability to use human tools (hand tools and vehicles). According to a 2012 document listing DARPA rules, robots needed to complete these tasks:

1. Drive a utility vehicle.
2. Walk across rubble.
3. Remove debris blocking an entryway.
4. Open a door and enter a building.
5. Climb a ladder and traverse an industrial walkway.
6. Use a tool to break through a concrete panel.
7. Locate and close a valve near a leaking pipe.
8. Replace a component, such as a cooling pump.

The entries were amazing, but many of the robots had difficulty maintaining basic balance, which you can see thanks to a compilation video* of robots falling during the competition. The robots fell. And fell. And folded. And collapsed. It was so extreme that engineers started joking that DARPA should change its name to DERPA.† It's funny, but pretty harsh. A lot of time, effort, money, and just plain brilliance went into creating these complex mechanical machines.‡ But think about it: Take a million dollars worth of tech, add a million dollars worth of labor by some of the top minds in engineering and robotics, and that's what you get—a robot that

---

*www.youtube.com/watch?v=7A_QPGcjrh0.

†For those of you outside the nerd loop, *derp* is a multifunctional word that can be a noun meaning goofy, silly, stupid, dorky (as in, "My brother just walked into a wall. What a derp.") or an exclamation used to describe one's own derpiness of the moment (as in "I called my best friend, but as soon as she answered, I forgot what I wanted to say. Derp!").

‡The winners, from South Korea, walked away with $2 million in prize money. Second place, from the Institute of Human & Machine Cognition in Pensacola, Florida, got $1 million; and third place, from Carnegie Mellon University in Pittsburgh, Pennsylvania, got $500,000.

falls over. Is that due to bad robot design? According to experts, hardly. Even the best, state-of-the-art technology can't replicate balanced two-legged walking, an activity that most humans have perfected by age four.

"The DARPA results demonstrate that we don't know so well how people manage to balance on two feet," says Dr. Manoj Srinivasan, associate professor in the Department of Mechanical Engineering at Ohio State University in Columbus. "There are still a lot of missing pieces. But even if you understand something well, there are limitations in how well you can build these devices and how well we can control them."

Dr. Srinivasan is one of those unsung geniuses working to understand more about what makes humans so good at maintaining balance while walking. Some of his most recent work involves studying normal human gait and designing robotic lower leg prostheses. He spoke with me about his research, which he hopes will one day contribute to helping robots *and people* walk with more stability.

Dr. Srinivasan explains that while building a robot that can walk twenty-five miles without stopping or falling down is not easy, it has been done. The robots that can do that are very simple—they are just legs, without knees, ankles, or even an upper body. All they can do is walk. DARPA-type robots are much more complex, and because they do more than walk (including climbing ladders and using tools), they have more body mass and moving parts. The engineers who work on these superbots have to not only track the leg situation but also be concerned with trunk position, arm placement and movement, hip motors, knee bends, ankle stability, and much, much more.

We take for granted that if our sense of balance is functioning well—if our inner ear, vision, proprioception, and sense of up are working—then we can remain upright and balanced while walking. It's not that easy. Do you remember Weebles? A Weeble is a children's toy shaped like an egg, with a weight in the base of the egg. No matter how you tip the Weeble, it always rolls back up, leading to the original tagline: "Weebles wobble, but they don't fall down." Egg-shaped with a rounded bottom and low center of gravity . . . *boom*—perfect balance. But people aren't shaped that way. First add in everything above your abdomen, which is the general location of an individual's center of gravity while standing still. Imagine a body that is just an egg, and then top it with a chest, shoulders, arms, and head. If the head moves backward, it alters the center of gravity just enough so the egg will rock backward a bit. Now imagine that the egg creature has a bendable waist, where the chest connects to the abdomen. Bending forward at the waist will cause the egg to become lopsided. Same thing with the arms—if they wave around like the arms of a high school cheerleader at the homecoming game, the egg will swivel and wobble as the position of the arms changes overall egg-balance. So you can see that even with an egg-bottom, the balance position is precarious and dependent on the movement and position of the other parts of the body. Now, finally, put that egg on stilts, because that's what our legs are like (to a Weeble, anyway). The upper body still changes the balance equation, but in addition, those stilt-legs have to balance everything.

This is the difficulty with making a robot that walks on two legs and also performs other intricate activities. You have to recalculate balance with every movement. It's a tangled knot of physics, mechanics, and programming.

"Roboticists are slowly making progress with walking robots," says Dr. Srinivasan, "Humans seem to be optimized for walking. If we could understand how humans walk much better, how humans control our legs much better, so that we can potentially build these machines better."

This is where kinesiology meets biomechanics.

One Small Step for Man, on a Treadmill

Human walking has been described as "controlled falling"—when we lift a foot to take a step, our bodies are in a state of physical uncertainty until the foot hits the ground. In that case, learning to walk is really about learning to control our bodies through that moment of falling.

Not true. At least not according to the experts I spoke with, and they have the data to back it up.

In 2014 Dr. Srinivasan and his colleague Dr. Yang Wang published a groundbreaking paper describing research that demonstrated that our strides aren't random, that where we put our feet as we walk can be predicted. There's a good chance that data from Dr. Srinivasan's research will be programmed into software that will power future robots, and it will help balance experts to see walking as something more than controlled falling. That's a "eureka" discovery to kinesiologists and bioengineers. Here's why:

The idea of controlled falling started as a simple way to describe our imperfect gait. Most of us don't think of ourselves as unsteady walkers. As long as we don't have a limp or other physical limitation, we tend to believe that we're very stable (so stable, in fact, that we don't typically think about our movements as we walk across the room to get a glass of water). But people are not perfect. Our muscles are not perfect, and the sensory information used to drive our muscles is not perfect. If we walk one hundred steps, each step will be slightly different from the one before as we continuously recalibrate our movements and take corrective action to maintain balance.

The idea of controlled falling implies that when we lift a foot to take a step, we are in a kind of micro-freefall and that the act of walking is a constant fight to control where we land. But kinesiologists believe the opposite is true, that we have control of leg movement, foot direction and placement, and balance. Yes, our gait is variable as we make minor tweaks in where our feet land, but that is to be expected from our imperfect sensory information and imperfect muscles. What others call falling, Dr. Srinivasan (among many others) calls "noise" in the system.

Noise in engineering doesn't mean the annoying sound coming from your neighbor's house on a Saturday morning. It has nothing to do your sense of hearing. Simply put, noise is the variation from idealized perfection. Since nothing in the world is perfect, there is noise everywhere, with everything that happens. The more factors that contribute to a given action, the greater the possibility of noise (since every factor has its own imperfections). For example, let's look at the act of grilling a steak. For the sake of this example, imagine that on Monday evening you grill a steak that sets the bar for perfection in every way. Your mission is to re-create that perfection every day for the next year. No matter how closely you try to match the original, perfect steak, you'll never match it. It's not because you're a bad or inattentive cook. Rather, there are factors and details that you have no control over. Steaks vary in thickness and fattiness, grill temperatures and flame patterns vary depending on whether you use charcoal or gas flames, seasoning varies with how heavy a hand you take to the salt shaker, and so on. Even if the steaks look identical to the naked eye, there will be tiny variations that make them different. All those variations from perfection are the noise in the system of grilling a steak.

When scientists measure gait, the noise is the variation from an imaginary perfect walking step. Our footfalls change from perfect as a way to correct for the noise in the system. What Drs. Srinivasan and Wang found was that the noise in walking—those deviations from a perfect gait—could be used to predict the next footfall. We are striding, not falling.

For this particular study, the researchers pasted eleven tiny reflective sensors on each participant—three on the pelvis and four on each foot. These weren't high-tech sensors but circular Styrofoam balls with reflective tape on them (the kind you might use to increase the visibility of your bicycle). Then the subjects walked on a treadmill surrounded by eight infrared cameras. When infrared light from the cameras hit the reflective markers, the light bounced back and was photographically captured as an infrared image. Images from the eight cameras were then used to reconstruct the relative positions of each sensor, creating a three-dimensional animation of a person walking. This process is called motion capture.* Simply watching the animation, while interesting and visually appealing, doesn't give researchers any useful information about dynamic foot position and placement. The movement differences happen in fractions of an inch, too small to be obvious. The real magic is done with mathematical equations, which allow researchers to see numerically what we cannot actually see with our naked eyes.

Dr. Srinivasan saw from the data that the body used noise to adjust gait. Specifically, he was able to predict 80 percent of foot placement based on analysis of noise. And the body didn't wait until it was "falling" during a step to make adjustments. Rather,

---

*This YouTube video made by Vicon, the company that made the cameras used by Dr. Srinivasan, shows how motion capture works in a laboratory: www.youtube.com /watch?v=sA3Kkq9kEiM.

Dr. Srinivasan could predict footfall as early as midstance, when the hip is directly over the opposite, weight-supporting foot. So it's not that we start to fall forward and then try to control the fall. Our brains and bodies calculate the effect of the noise—all the interceding factors—and make balance corrections well before the step is taken. You can't see it simply by watching a person walk; you need motion capture and sophisticated mathematics.

"It turns out that, with foot placement," says Dr. Srinivasan, "the corrective action is happening at the scale of a few millimeters and centimeters. If you just look at the motion itself, at the gross level, that subtlety is lost. Our visual perception is not sensitive enough to see [and register] those kinds of tiny little deviations. But those tiny deviations matter. They are the most important."

How important are those few millimeters or centimeters of correction made by your body? Without them, you would fall over. So it's not so much the controlled falling of lore, but fall prevention.

Dr. Srinivasan, working with PhD student Varun Joshi, also looked at what happened when there is an "external perturbation"—that is, the participants' walking path was "perturbed" by being pulled, pushed, or jostled. The goal was to see what happens to a person's gait when a physical force causes her to deviate from normal walking. Are we still able to integrate this additional noise into our mental calculations to predict future footfalls, as we do when we walk unperturbed? Or does an unexpected jostle turn our predictable steps into something more like a controlled fall? In this new study, participants started out by walking normally on the treadmill. As they walked, they were unexpectedly pulled to the side.* The results showed that our brains and bodies are really

---

*As I saw over and over in the course of my research for this book, scientists have to be creative and productive on a shoestring budget. For this study, I had expected that Dr. Srinivasan would have used some type of advanced technology to apply the

good at incorporating new and unexpected information. In fact the results were exactly the same as during unperturbed walking—study participants made the same foot-placement corrections, which were predictable before the correcting steps began. This time, however, the physical step-by-step corrections were more visibly obvious and could be seen without the benefit of computer calculations.

Chances are you have already experienced this type of perturbation on your own. Think about the last time you were jostled on the street. You're walking along, normal gait, when someone shoulders by and you are thrown off balance for a moment. You don't just keep walking with the same gait; you automatically adjust your foot placement so you don't fall over. That's the same principle Dr. Srinivasan was looking at, just magnified. You didn't have to think about what your feet or body should do to adjust to the sidewalk perturbation—you just did it. And while the jostle was unexpected, it probably didn't come close to pushing you off balance simply because your kinesthetic sense is always on guard to prevent an actual fall. Kinesiology is a critical part of our sense of balance. It helps the body know what to do in the event of a push. And that automatic analysis and correction is one reason why you are better at balancing than any million-dollar robot.

All this raises the question of what happens when those corrective mechanisms fail. How does the body respond then?

"You can imagine a few different things that can go wrong," says Dr. Srinivasan. "One is the imperfections that we're talking about that require corrective action, namely the noise in the muscles, the noise in the sensors, could be bigger for people with movement disorders or the elderly. Their ability to take corrective

---

perturbation. Nope. He had research assistants pull on a rope tied to a belt around the subject's waist. So, in this case, "external perturbation" = "yank on a rope." It makes me believe that scientists really need more research funding.

action might also be reduced. Depending on the particular population, different movement disorders might compromise different aspects of the system."

By examining the norms of foot placement and how the different types of noise affect balance, Dr. Srinivasan hopes that the information can be used in the control of robotic prosthetic devices or assistive exoskeletons.

"People are building these exoskeletons that reduce the effort of human walking, improve balance, or correct other issues with walking," he says. "They are getting better, but they are still far from perfect. We are now trying to figure out how best to control these devices. We're not talking about a legged robot that wants to walk around on its own—they [exoskeletons] need to work in concert with the human system."

Maybe I've led a sheltered life, but I didn't know exoskeletons were a real thing. I thought they were the stuff of science fiction and fantasy. The idea that there might actually be a device that could help people walk with greater balance and control is very exciting. I know so many awesome people hampered by movement disorders—Parkinson's disease, multiple sclerosis, muscular dystrophy—that I needed to learn more.

And that's how I got to meet Dr. Sawicki, the guy creating a real-life Iron Man.*

The Cool Area

I visit Dr. Sawicki in his basement lab at NC State. Outside his office are a few students scattered at open-space desks, and a whiteboard

---

*Dr. Sawicki would never make such a grandiose statement about himself. When pushed, he admits that he is working on a tiny piece of the beginning of a multiple-decades-long process of making anything even close to Iron Man. I'm still impressed.

with the handwritten message: WELCOME TO THE COOL AREA! As I would discover, it isn't hyperbole.

His lab is dominated by three pieces of equipment—a black metal staircase that climbs to a platform about halfway to the ceiling, a silver metal ramp that runs parallel to the staircase, and the biggest treadmill you've ever seen. The treadmill has two tracks, allowing measurements to be taken for each foot independently of the other, including heel strike, foot position, and weight. Any spare space is filled with tables, computers, monitors, and shelves that hold various components of the exoskeleton project. Motion capture cameras hang from the ceiling in a ring, all focusing on the treadmill.

Dr. Sawicki looks more low-tech than I had imagined. (But then again, I was imagining Tony Stark.*) With his bright yellow-and-orange plaid shirt, newsboy cap, and neatly groomed facial hair, Dr. Sawicki seems more art professor than human neuro-mechanics engineer. He is carrying a red insulated travel mug of coffee—it rarely leaves his hand.

As we enter the lab, I am introduced to two members of Dr. Sawicki's team: Dr. Tracy Giest, a postdoctoral research associate, and Richard Nuckols, a PhD student who is on track to complete his doctorate in 2017. They are all very welcoming, and they exhibit a comfortable camaraderie and visceral excitement as they talk about their work.

The scientists are very aware of the comparison with Iron Man, but they are quick to make sure I know that they are not staking any claim to that title.

"The devices that are most Iron Man–like tend to be real bulky," says Dr. Sawicki. "And because they're so massive, it's tough to

---

*I'm sorry. I'll try to minimize the Iron Man references. It's just that I'm a fangirl, and this research is so cool.

get good performance out of them, at least for the human side, for the user. What we try to focus on in my lab is what's called 'wearable robotics.'"

In other words, Dr. Sawicki's research focuses on the human physiological response to exoskeletons. The team spends less time developing the advanced technology and more time trying to understand how humans interact with "wearables"—how our muscles, tendons, and nerves respond to an attached piece of machinery and what the brain does with the different messages that come from the body when an exoskeleton is present.

Dr. Sawicki, Dr. Giest, and Richard Nuckols focus their research on the lower leg, especially the Achilles tendon, and the muscles and tendons of the calf. According to analysis, most of the energy for movement comes from the calf and Achilles tendon, so that part of the body is critically important for mobility in that it helps propel the body forward. Dr. Sawicki's interest in researching the role of this specific body part began at the University of Michigan, funded in part by grants from the Christopher and Dana Reeve Paralysis Foundation, where he was part of a team that built ankle exoskeletons to try to help people after spinal cord injury learn to walk on a treadmill with body weight support.*

One immediate benefit of the combination of weight support and exoskeleton devices is for physical therapy. The device, or robot, is attached to the leg to help apply the forces necessary to initiate leg movement, allowing physical therapists to focus on the high

---

*Dr. Sawicki's current lab is also set up for this kind of work. On the ceiling is a track that runs from the main door, over the metal staircase, and then over the treadmill. A person with spinal cord injury is put in a harness, which is then attached to the track by a connector that looks exactly like a clothing hanger. The harness can be adjusted to hold the person's entire body weight or any proportion of weight. So it is possible to hook up a person with no disability and adjust the harness to simulate what it would be like to walk on the moon or Mars.

level instruction rather than the basics of manually manipulating the leg.*

The challenge of building an exoskeleton for the calf isn't immediately obvious to someone without knowledge of body mechanics (like me), but Dr. Sawicki laid it out for me. When we walk normally (without an exoskeleton attached), our calf muscles and tendons alternately stretch and contract in a way that allows the body to convert kinetic and potential energy to and from those elastic tissues. As you push off to move forward, the stored elastic energy within the muscles and tendons is transferred back to the body's center of mass so you can take a step. So the challenge of building an exoskeleton is finding a way to generate that buildup of energy and the release of energy to power the body forward.

One successful early model used what are called "McKibben muscles," which are pneumatic artificial muscles made of bladders—basically surgical tubing with a little fiberglass or nylon weave around them. When you fill those tubes with compressed air, they bulge and shorten like skeletal muscle.

"What's cool about them," says Dr. Sawicki, "is that they are soft and flexible and easy to attach, so you can attach them around someone's leg in lots of different configurations pretty easily. What's not great about them is that the device requires a compressed air source, so either you have to plug into the building's compressed air, or you have to wear a scuba tank to walk around with them outside the lab."

Obviously not a convenient solution. And that's the Iron Man conundrum. How do you make a workable robot leg that

---

*Dr. Sawicki tells me that you don't realize how heavy a leg is until you are the one working with it to make it move. When he did this work, he could only work for a few minutes before becoming drenched in sweat and physically depleted. So the body weight support benefits both therapist and patient.

is portable, inexpensive, and lightweight? The weight is especially important because it affects your "gas mileage," or the amount of energy you use during an activity. For example, if you set out for a hike with a one-pound weight around your ankle, you'll be able to walk farther with less physical exertion than if you hike with ten-pound weights on your ankles. In other words, the less weight, the better your gas mileage. This is especially important for people who have a disorder that makes walking difficult, such as Parkinson's disease, spinal cord injury, or stroke. People with those disabilities are incredibly interested in improving their gas mileage, which is typically 50 to 70 percent worse than that of someone without a disorder.

Another potential consumer segment for exoskeletons might be . . . well, all of us. As we get older and our bodies age, we lose muscle mass and don't have the same physical capabilities we did when we were younger. So if you want to keep running, hiking, or walking into old age, you may eventually need mechanical help. An exoskeleton on the lower leg can restore power to a weakened calf and Achilles tendon. Or, if you are a healthy individual, an exoskeleton can harness the power from your calf and Achilles tendon and boost them up, or augment your abilities and improve your gas mileage.

"Or," Dr. Sawicki says, "we may give you some superhuman ability, like, you can accelerate faster and walk farther on the same amount of energy, have a higher vertical jump, things like that."

To that end, the progress of exoskeleton researchers is being watched and sometimes funded by the military, which sees promise in the possibility of having ground forces that can move farther and faster without increasing their energy needs.

So Close and Yet So Far

In his search for the answer to the Iron Man conundrum, Dr. Sawicki worked in a muscle physiology lab to try to understand how muscles and tendons work together, with an eye to bringing them back to exoskeleton design. Basically he was trying to get what he calls "bio-inspiration" from the human system. He learned that the human calf and Achilles tendon act kind of like a clutch and spring, or a little like a catapult.

"It's not the muscles themselves driving most of the motion," Dr. Sawicki says. "It's actually the tendon, which stretches and then recoils, kind of like a rubber band. And it's that recoil of the rubber band that really propels you through the world. More than half of the energy that you use to move around is coming from the snapback of the rubber band in your ankle. That's pretty cool."

Springs are lightweight, and they don't use any energy; they're just material. And you don't have to worry about an external power source like a motor when you have springs. You just have to find a way to latch on to the spring and then release it at the right times. So the team created a device that uses this spring-and-clutch concept in a form that can be adjusted and worn by just about anyone. I had a chance to see this experimental exoskeleton in the lab. The device feels very light, about the weight of an average hiking boot. Because of its simplicity, it looks a bit like something that might be part of a homemade Halloween costume. There's an average aqua-and-neon-green women's sneaker at the base. Attached to that is a carbon fiber shank frame, which holds the device in place, and at the back is—as promised—a spring and clutch. The spring looks no different than the spring on your porch screen door, and it functions pretty much the same. There are also a ton of wires coming from it, but those are there to take measurements for research.

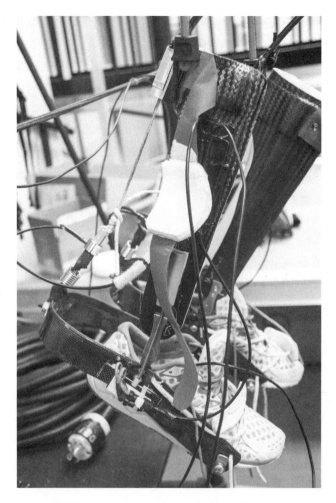

Experimental exoskeletons, with sneaker base.
Note the spring mechanism running up the back.
Carol Svec

This model was tested on healthy young people, who provide baseline measures of efficacy and efficiency. The test subjects walked on the treadmill, while the ring of cameras provided motion capture imagery. The wired-up device gave readings of ankle muscle activity, and gas mileage was calculated based on measurements of whole-body oxygen consumption and carbon dioxide production. The researchers found that by using this simple spring-and-clutch design, walkers improved their gas mileage by about 7 percent. That's a real improvement for anyone, but especially for those with disability or extreme-fatigue jobs.

Dr. Sawicki's team also works with people who have had strokes and still have some physical deficit. A stroke usually affects one side of the brain, limiting or destroying function on just one side of the body. Someone who has suffered a stroke can't properly use his or her calf on one side of the body to push off and therefore starts to rely more heavily on other muscles, especially those around the hip joint. But those muscles use a lot more energy. Dr. Sawicki's idea is to apply the spring-and-clutch concept to an external device that can help stroke survivors with the push-off phase of walking. Once push off is normalized, it should be possible to restore the symmetry in their gait. No more limping!

So far, stroke survivors have offered mixed reviews of the exo-skeleton. Some of them love it. On the treadmill, they walk longer and faster than they have since their stroke.

"One guy said that, for the first time in a long time, he feels like he has spring in his step again," says Dr. Giest.

That's exactly the idea—use a spring to get the spring back.

However, some people don't like using the exoskeleton. They have gotten used to their limp, and the extra boost makes them feel out of control and unbalanced. Part of the reason some stroke survivors won't walk faster is because they are afraid of falling. With the boost of spring energy, these walkers feel that the device is too dangerous for them.

Even if an Iron Man suit existed, you couldn't just slap it on different people and expect it to function the same way for every wearer. People are complex biological creatures that adapt to changes at different rates, depending on the change and the individual. Plus, it turns out that the individual results of strokes are as different as fingerprints.

"The field is at a point where I think we need much better ways of understanding individual deficits poststroke," says Dr. Sawicki, "so we can then prescribe the appropriate technology or therapy. You can't just override the impaired underlying human system with the robot and expect things to go perfectly. That has actually been a real challenge with exoskeletons in general, that the engineer in us wants to just build something, then just slap it on someone and expect the human to behave the same every time. That just doesn't happen. We're very early in our understanding of how to attach these things."

Muscle synergies pose one of the more intricate and complex problems of deciphering the physiologic needs of someone post-stroke. Synergies are those groupings of muscles that are wired together by a neurological connection to the brain. In walking, there might be a synergy for pushing forward or for having an affinity for turning left. Some of those synergies might have been destroyed by the stroke, and some of the motor memories might have morphed into other synergies, causing you to lose your flexibility. Nuance in movement may be lost. Part of the challenge for people in this field of research is figuring out how to help people regain that nuance.

"If you could figure out what a person's synergy pattern looked like poststroke," says Dr. Sawicki, "then you can use that pattern to design a robot well. That, I think, is a really promising approach—an individual prescription based on their neural or motor control footprint, and also their biomechanical footprint. I think that's where the field is headed. We're not quite there yet."

By the end of my visit, I am charged up with the possibilities of exoskeletons. My stepdad is someone whose movements have been severely limited by Parkinson's disease. Many people only know Parkinson's as the disorder that causes you to shake and have uncontrolled movements. That's true. But one of the lesser-known symptoms is difficulty initiating walking. It's like the mind tells the feet to move, but they don't respond. There are days when my stepdad says, "I just can't get my feet to go." I imagine a time when we could strap a pair of Dr. Sawicki's calf-high exoskeletons on him so he could walk—just *walk*—the way he wants to. But then I think how he has difficulty with balance even now. What would he be like with boosters on his sneakers?

Dr. Sawicki sees that problem, too, and acknowledges that the field is in its infancy. Right now, the technology of movement and the science of balance are two separate arms of research that will need to come together at some point in the creation of the ideal exoskeleton. The creation of Iron Man.

"We're way more focused on the technology side," says Dr. Sawicki. "You know who you should talk to? Jason Franz and Lena Ting. They are all about balance."

And so I did.

13

With Luck, We All Get Old

Fall Prevention

WHEN I WAS ABOUT TEN, my grandmother would often ask me to thread needles for her because she couldn't see the tiny eye of the needle without her glasses. That's when I started noticing that a lot of adults walked around with glasses on top of their heads or strung around their necks on lanyards. I remember vowing that I would never let my eyes get so bad that I needed help reading the label on a soup can. I was different, I thought. I would never get what my childhood self used to call "old people glasses."

Then I turned forty-two. Seemingly overnight, the unimaginable happened, and I became one of the "old people." These days, my close-up vision seems to get a little worse every year. This year, my eyes decided that things weren't complicated enough, so now I have my everyday "need them to see" glasses, reading glasses, *and* computer glasses, because the working distance of my laptop is different from the distance I hold a book when I'm reading. So much for childhood vows.

I tell that story because one thing I've learned in the course of my research for this book, one bit of information mentioned and stressed by nearly all the experts I consulted, is that our ability to balance gets worse as we age. And (with luck) we will all get old. Just as our eyes physically change so that everybody—whether they like it or not—ends up needing reading glasses, all our balance systems degrade over time. The vestibular system, vision, and proprioception all get weaker. Even our automatic body reactions slow down, so that we are less able to stop a fall after we temporarily lose our footing. There are things you can do to compensate for this age-related decline, but you cannot prevent it from happening. Even if you exercise and eat right, it will still happen to you. You will get old. You will walk more slowly. Your balance will get worse. And you will probably fall.

Understand the Enemy

As I wait for the virtual reality experience to begin,* I get that same surge of excitement I felt in other VR devices, but this time there's something different: a treadmill.

I'm in the lab of Dr. Jason Franz, assistant professor and director of the Applied Biomechanics Laboratory, part of the Joint Department of Biomedical Engineering at the University of North Carolina at Chapel Hill (UNC) and North Carolina State University (NCSU). Like most of the research labs I have visited, the room is sparsely decorated, with oatmeal-colored walls, a linoleum floor, a standard T-bar drop ceiling, and fluorescent lights. There are several large pieces of equipment, but the star of the room is the

---

*Yes, we're back to virtual reality, but this time it is the tool, not the focus.

floor-to-ceiling semicircular movie screen placed so that it curves around the front of an extra-large treadmill. The treadmill has two tracks—one for each foot—and it has the ability to measure ground forces during walking.* Motion capture cameras are suspended from the ceiling all around the VR treadmill.

Dr. Franz is in his midthirties, but he has almost a chameleonlike quality—when he discusses the details of his research, he carries himself in a way that makes him seem older, but when he laughs and talks casually, he appears much younger. Maybe that's because of his brownish-red hair and freckles, which seem to be standard TV and movie typecasting features for plucky or impish child actors.† I'm always a little nervous when I go to visit scientists. I'm in awe of their intellect, and I worry that I won't be able to keep up with their flow of ideas. But Dr. Franz's manner is so warm and inclusive that he makes me feel immediately at ease.

To give me an idea of how one of his research projects works, he tells me to hold on to the front bar while he starts up the treadmill. This ensures that I won't fall when the belts start moving, but it also puts me in the right position to get the full VR effect. The screen extends from my feet to the ceiling and in an arc around me, from my left hand to my right. Having a full 180-degree screen means that the projection spans my whole visual field, central focus and peripheral vision, making this an immersive VR experience.

You may recall from chapter 5 a discussion about optic flow, the perception of motion as objects move past you in a continuous stream. The most common everyday example happens when you are in a car, watching the scenery move across your field of

---

*This is the same type of treadmill used by Dr. Greg Sawicki for his research with exoskeletons (see chapter 12).

†I think I have already established that I consume too much visual entertainment.

vision as you drive by. The effect can be simulated on a screen by showing objects that move in that same patterned way, such that an object in the center of the screen looks small but grows larger as it moves to the edges of the screen. The most common example is the old Microsoft star field screensaver, which made you feel as though you were moving forward through starry space.

The VR projection used in Dr. Franz's research is an optic flow pattern designed to look like you are walking down a never-ending hallway. It is not a fully animated scene—there is no actual floor, no walls, no ceiling. There are just squares of light that are arranged to suggest a hallway, along with rectangular shapes that depict side doors and windows. It doesn't sound like much, but the effect is a powerful trick, fully convincing the eye and brain that there is a hallway. As I walk on the treadmill, it looks *and feels* as though I am walking down that dark, spaced-out hallway.

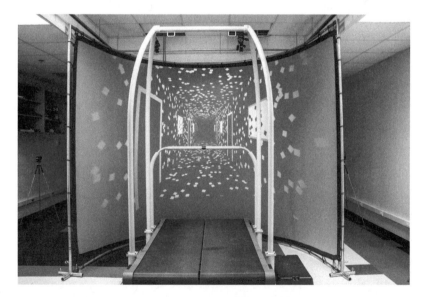

Treadmill (note the double belt) and curved projection screen.
Carol Svec

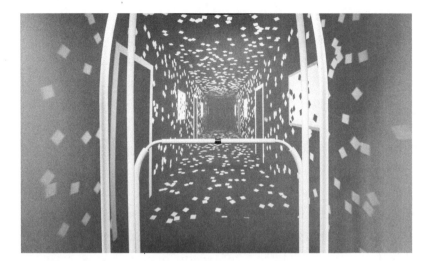

Close-up of the virtual hallway.
Carol Svec

I'm enjoying the VR stroll when Dr. Franz makes an adjustment and that hallway, instead of just moving past me like a normal
hallway, begins to move side to side. The movement is very slow
but at frequencies that my brain interprets as, *Oh, no! I'm falling
to the side!* And I can feel my gait and body change in an effort
to prevent the fall. Remember—the only thing that changed was
the movie on the screen in front of me. The ground is stable, and
the treadmill is moving at the same speed as before. My vestibular
system, proprioception, and gravity sensors are all relaying a steady
message because nothing is *really* happening. But my vision is
getting a new signal that I'm falling, and as a result, my feet move
out to a wider stance to give myself a firmer base, and I can feel
my body tense up a bit.

Like with all really good perceptual illusions, I can't help but
react the way I do. I knew in advance what was going to happen,
and I knew what actions were expected. I set my mind to not react,

to understand that there was no actual tilting of the room, and to simply maintain the same walk no matter what happened. But I couldn't. My legs took on the wider stance automatically.

Adding the sway to the virtual hallway is what scientists call a visual perturbation. This is similar to the regular perturbation (or interference) used by Dr. Manoj Srinivasan in his research,* except he physically yanked at his subjects to disrupt their gait. Dr. Franz alters gait merely by changing the on-screen projection. There's not enough sensory conflict to cause motion sickness, but just enough to make me feel askew and adjust my stance and gait accordingly.

"It's just instinctive," says Dr. Franz. "You take really wide steps, short steps. Your body goes into a sort of conservative balance-control strategy. We think people alter their movement patterns to synchronize their postural control to match the visual feedback, which then requires really strong corrective motor responses in the form of foot placement adjustments from step to step."

After a few minutes, I can feel my gait returning to normal even though the hallway is still swaying side to side. It's like my body understands that I'm not going to fall. I'm not consciously aware of the adaptation—I am busy chatting away and eventually notice that I am no longer walking like a linebacker.

Dr. Franz and his team stumbled upon this particular line of research while looking for a way to measure balance differences between young and old people. We know there are balance deficits that happen with aging, but in order to scientifically analyze the differences, you have to be able to find and measure them. You can't just wait until somebody falls and then backtrack to determine the cause. Dr. Franz and company conducted a big balance intervention study, with just standard, conventional

---

*See chapter 12.

standing-balance and walking-balance testing and training. The problem was that the team needed some outcome, something they could measure that would tell them whether the balance training had any effect.

"We were doing all the conventional stuff," says Dr. Franz, "and we thought, 'Let's throw something else in the mix; let's do a couple perturbations.' So we had [the research participants] walk on a balance beam, we did some cognitive testing, and then we wanted a sensory perturbation."

That's when things got interesting.

One of These Things Is Not Like the Other

Whether they were walking on a balance beam, doing complex mathematics,* or walking on a treadmill in front of a VR screen, young subjects and old subjects performed exactly the same. And then Dr. Franz and his team turned on the visual perturbation and saw age-related differences emerge that were not otherwise apparent.

"The old† subjects could barely walk and stand straight at the same time," says Dr. Franz. "It was a really rich response that we didn't expect."

---

*I can't imagine being able to balance and do complex mathematics at the same time. Where did these multitalented research subjects come from? I just about fall out of the chair when I'm trying to balance my checkbook.

†As a member of the baby boomer generation, I instinctively cringe when using the word *old*. It feels like a pejorative, even though it is simply a descriptor of a life metric. We all have a height, we have a weight, and we have an age. They are numbers, not insults. At five feet one, I fit everybody's category of *short*. It makes no sense for me to rebel against being part of that group; my height is simply a number. Like age. Scientists typically label the over sixty-five crowd as *old*. They (and I) acknowledge that some people are a more youthful sixty-five. I know many of them. But, like needing reading glasses, some physical changes happen naturally as a body ages, and age sixty-five seems to be a turning point. I felt the need to say this because I feel bad and guilty every time I write the word *old*. Does it help if I blame it on the establishment?

Why would this happen? Dr. Franz theorizes that it has to do with which portion of our complex balance system is dominant. Young adults prioritize the physical, proprioceptive/somatosensory feedback—that *pressure* of body weight on their feet, that *force* information from under the feet—to guide their motor responses. That's the information they use to plan and execute balance from step to step. In aging, physical sensory systems are among the first things to go. The decline there is faster than is seen in some of the other sensory systems. Instead, older people have much more visual reliance. So even though you lose visual acuity as you get older, you actually have a greater reliance on vision as a reference for balance control. As Dr. Franz puts it, visual reliance "goes through the roof."* In that case, it's easy to see why the visual perturbation (swaying hallway) would have such a dramatically different effect on young and old people.

As a relatively young person walking through the VR swaying hallway, I had the experience of gait change, but I barely felt the change. It's amazing that a small trick of the eye could make people fall over simply because they are older. This discovery is so powerful that this sensory perturbation test might someday be used as a way to identify who might be at risk of falling *before they fall.* That italicized phrase is important because with the current process, a person isn't considered to be at risk of falling until they have already had at least one fall. If you go to a doctor's office or clinic these days, the conventional test is to ask, "Have you fallen?" If yes, then you might be sent for further tests or for physical therapy. But if you haven't yet fallen, you aren't considered at risk for a fall—you're not a "faller." But every faller had to have a first fall,

---

*Remember that *All About Eve* reference in chapter 6? It works here, too. This is another time when vision steps in to take up the slack when our more primary balance sense begins to fail.

and that first fall can sometimes result in serious injury or even death. There's an opportunity for this simple visual test to be used as a way to target at-risk individuals prefall. It has the possibility to revolutionize early fall prevention.

These results should also serve as a public service warning to people over age sixty-five, that, like it or not, your greater visual reliance with age generally makes you at greater risk. It doesn't matter if you run regularly or can walk a balance beam; if something in your environment suggests movement to your visual system, you may fall—and you won't be able to control it. If you read chapter 6 on persistent postural-perceptual dizziness (PPPD), you know how many things might trick your eye into seeing movement where there is none. It's no wonder so many people fall.

The good news is that, in addition to its future use in diagnosing the risk of falls, the VR swaying hallway might also find a place in balance rehabilitation.

"We think that if you can push people that need to improve their balance," says Dr. Franz, "if you can nudge them to begin to practice responding to the virtual perturbation, you might be priming their system to respond more effectively if something unexpected happens in their [real] environment."

This has been shown to be true with physical perturbation-based balance training. Researchers at the University of Toronto have found improved balance measures after training with a motion platform that moved unpredictably (that was the perturbation). And a review of research found that people who participated in one of those physical perturbation programs were less likely to report falling. That's exciting, and cause for optimism when it comes to the future of VR-based perturbation training for balance.

"Virtual reality is very exciting," says Dr. Franz. "That and telerehabilitation. They go hand in hand, so that if you need to rehab someone, you can do so remotely. All of their information, the metrics of their course of rehab, their level of practice and performance, can be taken and stored in the cloud, and clinicians can monitor all of that. Virtual reality and telerehab are cutting-edge for sure."

And maybe, Dr. Franz tells me, we need something cutting-edge, something different. A commonly cited statistic from the Centers for Disease Control and Prevention (CDC) states that falls affect one in four people age sixty-five or older each year. Falls are the leading cause of fatal and nonfatal injuries for that population of older adults. But that's not the whole story.

"Since 2001," says Dr. Franz, "there has been an increase of 46 percent in the number of emergency department visits for falls. And you might think, well, sure, that population is growing. But the population of people over age sixty-five has only increased by 17 percent. There's two ways to interpret that. One way is that, well, when people fall, they are just going to the ER more, so that might explain the increase in ER visits. The other is that, despite our best efforts, we're actually doing a worse job now preventing falls than we were in 2001."

That's why the type of work Dr. Franz and his team are doing to help identify people at risk of falling *before* they fall is so critically, life-savingly important. And why we might need to look at new technology—like virtual reality.

Kayaking Without Getting Wet

As I discovered several times over, virtual reality is powerful to the senses. It has the ability to draw you in and take you into other worlds. With my little Google Cardboard headset, I have experienced what it is like to dive with sharks, ride on one of the world's tallest roller coasters, and walk with dinosaurs. I would have been very happy to spend a day in the Duke University's DiVE cube, walking through a French cathedral or flying through an abstract world of modernistic shapes. This technology totally fools my senses. And in Dr. Franz's VR, I discovered that my vision is so totally fooled by the experience that my gait changed automatically.

Some researchers are beginning to integrate VR technology into programs designed to improve balance. The idea is that the people who need balance training the most (i.e., older people) are also most likely to be afraid of falling. But what if you could put them in an environment where there was absolutely no risk of falling but they could still train their muscles in a way that helped them improve balance? And, to put icing on that exercise cake, what if you could also make the physical therapy regimen feel less like a tedious homework assignment and more like a reward?* I mean no disrespect by that. As the saying goes, some of my best friends are physical therapists. Physical therapy has helped me many times over the years after surgery or injury. But physical therapy is *hard*. At best, seeing the therapist is uncomfortable, and at worst, it is through-the-roof, I-want-to-report-you-for-abuse painful. It is difficult to find the motivation to do the exercises at home. Virtual reality seems like the perfect solution.

Several research groups are looking at various VR programs to help older adults with balance problems. For example, one group in South Korea developed a 3-D virtual reality kayak program

---

*Remember how popular the Wii *Sports* games were? People who were either unable or unwilling to actually play tennis, go bowling, or take up fly-fishing found themselves entranced for hours. Generations of families gathered in the living room to golf together. VR is just a step more "real" visually, so programs that can be made fun should be very successful.

designed to improve cognitive function, muscle strength, and balance in older people (average age about seventy-three). The participants sat in seats designed to feel like kayaks, and they were mounted on springs to simulate the roll and resistance feel of water. They paddled as they watched a projected 3-D movie of kayakers on a river, from the perspective of being among them.

Kayaking is kind of perfect for all-age balance training. The paddle motion is relatively easy, compared with canoeing, and being seated allows a greater number of older adults to participate. And, as anyone who has ever kayaked can tell you, balance is the most challenging part of the sport. Kayaks are narrow, and you need to adjust and readjust your body weight distribution with each dip of the oar. The risk of tipping over is high. It's great fun and very relaxing once you get the hang of it, but I can't imagine many seventy- or eighty-year-olds feeling safe while trying it for the first time. That's the benefit of VR—challenging experiences you can have in your own home. And after six weeks of two kayaking sessions per week, participants demonstrated actual improvements in both sitting and standing balance.*

Balance Meets Exoskeleton

Of all the changes that happen as we get older, one of the most visible and individually frustrating is decline in walking speed and balance. We have three main calf muscles that power walking, and they do so while consuming very little oxygen, making it very economical power. In aging, we don't use those muscles as much, for some reason. We compensate by using other muscles—quadriceps

---

*As might be expected, the 3-D movie caused dizziness in some participants, but as I learned from my adventures in VR worlds, that symptom usually decreases with more experience.

and hip extensor muscles—and those muscles aren't architecturally designed to provide the same economical locomotion as the calf muscles. Scientists believe that's one reason why older people take shorter steps and walk more slowly.

"For an old adult to travel from point A to point B," says Dr. Franz, "regardless of how long it takes them to get there, older adults will walk from point A to point B and consume 8 to 31 percent more oxygen. It's very tiring to walk just the way they move."

The major part of Dr. Franz's work involves trying to find the answer to why older adults walk slower, and why they consume more oxygen when they do so. The simplest explanation—that all muscles get weaker with age—doesn't cover it. If you do strength training or are just generally physically fit, you don't get a benefit that translates to more economical walking. You get stronger, sure, but you won't change the way you move. One study from researchers in Norway found that seniors who were generally active, with a sports-heavy background, were no better off than sedentary seniors when it came to balance.

But there has been a breakthrough. In the past year, Dr. Franz and his team used ultrasound imaging to look under the skin. Those three major calf muscles attach to the Achilles tendon. For many, many years, the Achilles has been thought of as one tendon, but it turns out that there are three separate tendon bundles, one from each of the three calf muscles. Those tendon bundles merge together and become the Achilles tendon. The researchers discovered that even though the three tendons are connected in this way, those bundles move independently and slide against one another— in young adults. The same ultrasound imaging in eighty-year-olds doesn't show any of that sliding movement. And researchers have correlated the sliding behavior with the mechanical output of the calf muscle, so more sliding means better power output from those muscles during walking. It sounds like this sliding motion is a good thing, but the discovery is so new that no one really knows what it all means. How do tendon changes affect muscle changes? How would those changes affect joint function? And how would that affect overall movement during walking?

"The NIH has funded us for the next five years to figure out what's going on," says Dr. Franz. "We think that adhesions* develop between the bundles [with aging]. That unfavorably affects the muscle behavior, and that is one reason why older adults use those muscles less and compensate in other places."†

And because Dr. Franz isn't busy enough, he is also planning a collaboration with Dr. Greg Sawicki, the scientist introduced in chapter 12 (a.k.a. the guy creating Iron Man) to use the calf muscle advanced imaging techniques to help tailor assistive technologies (such as exoskeletons) for older adults. The idea is that, say, an eighty-year-old man would go to the Franz lab to get a complete characterization of the biological tissue structure of his calf. Then the information would be transferred to the Sawicki lab, where they would build an exoskeleton that could, conceivably, restore the function of the calf to what it was when the man was twenty.

This would be a bold step into science fiction–level coolness. It could be one way to turn back the clock on the physical disabilities of aging, allowing us to walk faster and steadier and maybe helping to reduce some of the contributions to that awful statistic on falls.

---

*Adhesions are growths or connections between body parts that should not be connected. They are often described as internal scars. I imagine adhesions as well-chewed gum that gunks up the works and then dries hard in place.

†Dr. Franz wonders about a lifelong history of running, since it is a very tendon-driven activity. There is stretching and recoiling of tendon with each step. Might that prevent the onset of adhesions? It's possible. Maybe. A whole lot more research is needed before the question can be answered.

Virtual reality, ultrasound imaging, and exoskeletons. Drs. Franz and Sawicki are working on the sharpest point of that cutting edge of research on aging.

What Does This Mean for Us?

Because of Dr. Franz's work to promote balance in older adults, I asked him if his work offered him any special insights that he would like to share. What would he tell his mother or grandfather to help them stay upright?

"I'm always nervous to answer this question," says Dr. Franz. "We have no idea if older adults are ready to accommodate some possible changes, like increasing walking speed. Maybe there's a reason why they choose to walk slower, and intervening at all might be a bad thing. We think that there's a reason we should intervene in the gaits of older people—it costs them more metabolically to move, and so they get tired out more, they can't keep up with their grandkids. So there probably are functional reasons why we should intervene. But if we do so, are there negative consequences that might emerge that are completely unanticipated? I don't know."

I posed that same question to many researchers, and they all had different versions of the same answer. While research is beginning to peel back the layers to see what goes on inside the bodies of older adults, we don't yet know how to advise people to prepare for the balance challenges of aging. There are so many different systems and factors involved, and as we have seen, making a change in one system could offset balance in an entirely new way.

One bit of advice that I first heard over a decade ago and which is still a basic component of balance instruction is that we should be more active in our younger years, gaining experience in a variety of sports and general life activities. The thought has been that a well-trained body might provide some benefit—if not outright defense—against the ravages of aging.

Even that well-worn counsel may not be true.

To discuss how to prepare our muscles for the aging battle, I spoke with Dr. Lena H. Ting, professor of biomedical engineering at Emory University and Georgia Tech, and also professor of rehabilitation medicine in the Division of Physical Therapy at Emory University in Atlanta, Georgia. I asked her whether it was true that we should try to build a repertoire of activities so that we have more of our "muscle library" to call upon later.

"It certainly is a hypothesis we would like to test," says Dr. Ting. "I actually had a call from a colleague last week asking about whether that has been demonstrated. I don't think it's been demonstrated scientifically. But it certainly needs to be explored. It's difficult to test."

The reason it is difficult to test has to do with muscle synergies or muscle modules, which are the collections of muscles involved in a task, all activated by a single neural pathway. Consider that we have a couple hundred muscles, and each muscle doesn't correspond to just one movement. A single movement of your leg will involve your joints, as well as muscles in your arms, your trunk, and the other leg, all interconnected in synergies. For each type of movement you might want to do—dancing, kicking a ball, or walking a tightrope—you need to learn the pattern of coordinating all those muscles together to achieve that task. By building up a library of these subactions, you can generate a wide range of movements.

Here's the thing, though: the muscles you use to move your leg are similar to somebody else's leg-move pattern because humans have structural similarities in our legs and in our environment, but the particular leg-move synergies *you* create will vary somewhat from the synergies of other people. Researchers are just beginning to look at how older people and younger people differ in muscle use and synergies.

One general observation is that older adults tend to coactivate their muscles more for balance control and for gait, perhaps to compensate for feeling unstable. This means that, instead of specifically activating just that one leg-move muscle module, older people involve more muscles. So the muscles are overly active, sometimes even fighting each other, but they get the movement done. The cost, however, is that movement takes more energy. Again, we see that the gas mileage of muscle use for older people isn't great.

If gas mileage were the only problem, then these coactivating muscles wouldn't be such a consideration. But they also make us unstable and more likely to fall.

"What I think happens is they [older people] feel unstable," says Dr. Ting, "and then they coactivate their muscles. It's like when you learn to ski—at first it's very tiring and you're activating all your muscles because you don't really know how to move. With all those activated muscles, you become sort of stiff, and then you fall over sooner. As people get more proficient at skiing, or any movement, then they can relax and only activate the muscles they have to."

This makes sense. With all the balance systems degrading with time, it's only logical that older people trust their bodies less. They tense up muscles inappropriately, making them more rigid and more likely to fall. These muscle coactivations aren't necessarily conscious; it's one of those ways that the nervous system controls the body to perform movements. So really, our body's compensation for feeling unstable can actually increase instability. Another conundrum.

"Balance control is something that we almost never think about," says Dr. Ting. "In fact, I thought it was really boring as a graduate student. You know, nothing moves. It's not something that most of us are consciously aware of until we start to lose that ability. And then when you do lose it, it becomes very evident how complex it is, how much it relies on a wide variety of sensory systems, integrating them, and very specific feedback control to maintain upright posture, as well as a lot of different strategies that you could use to activate your muscles."

One takeaway message from Dr. Ting's work is that we should, in general, be as active as possible. The nervous system has what is called use-dependent plasticity. In other words, use it or lose it. The more you activate parts of your nervous system, the stronger the nerve-to-nerve connections. And the stronger the connections, the more easily you can tap into them when needed. One activity I know I have lost is the ability to do cartwheels. I used to cartwheel everywhere. I could do it without thinking about it. But after years of not using it, I lost it. Today I wouldn't even know how to start a cartwheel—which hand goes first? But my friend Joy, who used to be a gymnast, can knock out cartwheels down the aisle of a grocery store, even decades after her last competition. Her cartwheel wiring is very strong. So, as your body changes with aging, you need lots of different strategies to compensate for the decline in balance systems. If you don't have a history of activity that built strong wiring, then should your balance falter due to age, stroke, or illness, you'll have to go to rehab to learn how to move again. Better to work on that wiring early, when you're stronger and healthier. Dr. Ting recommends any kind of physical activity, as long as people aren't injuring themselves—walking, dancing, and tai chi are good whole-body activities a person can do with few negative effects.

"We can't prove that people with more synergies would be more resistant to motor deterioration," says Dr. Ting, "but we do anecdotally know that a lot of people who participate in our research studies are very healthy, and so we always do these surveys of how much activity they do. We have amazing cases of people who are playing tennis and volleyball in their seventies, and they're doing great. I think that does contribute a lot to their health and their ability to compensate for the inevitable degradation of sensory function that will impair people's balance as they age."

Just a Butterfly Touch

Another fascinating research area Dr. Ting is working on concerns what scientists call "light touch" or "fingertip touch." This phenomenon was originally discovered by Dr. John Jeka and Dr. James Lackner back in the early 1990s. You can demonstrate the basic premise of light touch in your own home. Stand within comfortable reaching distance of a table and try to balance on one leg. (Put both feet down if you find yourself at risk of falling—no accidents on my watch!) Notice how wobbly you are as your body tries to stabilize itself. Now repeat the experiment, but this time lightly place an index finger on the table. Don't press or lean on the finger, just place it lightly on the surface. The touch will help you stabilize. Your balance improves, almost like magic!

The idea behind light touch is that your index finger isn't applying enough force to mechanically stabilize you. Rather, it provides a sensory reference that is more reliable than your wobbly leg. Just that one, small sensory touch is enough to help your body "understand" stability. And it happens in a split second, literally. A study published in 2005 found that it takes just a half second for your body to begin balance adjustment based on the fingertip information.

This is a potent argument for using handrails, a cane, or a walker, even if you're not supporting your weight on them. Many people resist using a cane because they don't feel their balance is bad enough to require that kind of support. But the theory of light touch says that the cane can provide that kind of grounding sensory information, even if you're using it more like a prop.

"Without that additional sensory information," says Dr. Ting, "people have to get the information from their vestibular system, which frankly doesn't work very well if you are standing quietly; or through sensors in their feet, which degrade with age; or through proprioception of the ankle, which is a very small signal compared to the larger movements of the upper body. But if they can reach out and touch something, [they] can sense that motion better."

This shows how just a little bit of information can be the absorbed and used by the body to the benefit of your balance. Researchers, including Dr. Ting, are looking into real-world applications for the phenomenon. How might assistive devices be designed, or what kinds of improvements can we make to our homes that would allow us to take advantage of the powers of light touch? And it gets even better than that. Dr. Ting and her colleagues are planning to design a robot that could physically interact with a person through light touch—the same type of light touch you might have when you and a really good dance partner are gliding through a waltz. This type of robot would help in reha-bilitation therapy, assessment, and even possible fall-risk diagnosis. A pilot study is already being used to evaluate the performance of a prototype of a dancing therapy robot. When expert dancers put the bot through its paces, half of the dancers thought the robot was a good dance partner and that it was a good follower. Personally I don't care how good a dancer the robot is—the future of balance rehabilitation sounds like a lot of fun.

14

Balance Cycles Around

Coming Full Circle

IF YOU WANT TO SEE REAL-WORLD research in action, visit the Toronto Rehabilitation Institute—University Health Network, the world's number-one adult rehabilitation research institute. I had a chance to tour the facility with two of its (many) research stars—Dr. Behrang Keshavarz and Dr. Jennifer Campos, who is chief scientist for the Challenging Environment Assessment Lab (CEAL) at Toronto Rehab and assistant professor in the psychology department at the University of Toronto. I can report that the institute puts all balance research into perspective. Yes, there are some astonishing virtual reality research rooms, but what really makes the institute stand out is its dedication to providing environmental solutions to balance problems.

"What's really important to us," says Dr. Campos, "what I think distinguishes us is that we very much prioritize the real world implications of our research. We want to be able to change policies, to change practices, and create products that directly benefit individuals."

It's a mission that they take seriously. For example: their staircase success story. Researchers, led by Dr. Allison Novak, set out to optimize the design of staircases to reduce falls. They spent years changing the dimensions of staircases, analyzing the kinematics of people's behaviors when using the stairs, and looking at which set of dimensions minimized fall risk. They discovered that the run length—the depth of each step—required for residential homes in Canada was smaller than optimal. Results from their laboratory research were presented to policy makers and the government, resulting in a positive change to the national building code to increase the run length by two inches for each step. Now the longer run length is the standard for all residential buildings. That two-inch change will help prevent falls, potentially saving lives and millions of dollars in healthcare costs.

It doesn't get more real world than that.

Similar research is being done in several of the institute's labs, including HomeLab, which is quite visually impressive. It looks like a tricked-out single-story home, designed to mimic a generic type of environment older adults might experience on a daily basis. The purpose is to see how people interact within the home/lab, address balance issues, and come up with new technologies to allow people to age in their own homes, living independently and safely, for as long as they can. That's the ultimate goal.

For example, researchers are testing a fall-detection system, which uses "computer vision"—a camera that looks like a smoke detector mounted on the ceiling can monitor the living area and detect if someone has fallen. If it detects a fall, the computer starts

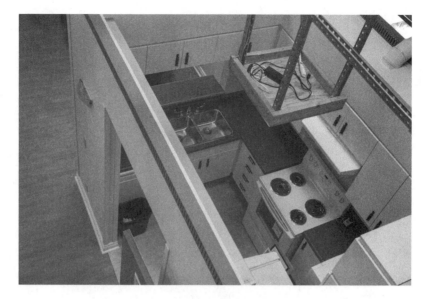

Overhead view of the kitchen space in HomeLab.
Carol Svec

a conversation, asking (for example), "Mr. Smith, are you OK?" If there's no reply, the computer calls emergency services. If there is a reply, the computer will go through a sequence of options, such as, "Would you like me to call your daughter, Sally? Would you like me to call your neighbor, Paul?" Then the computer can initiate a phone conversation or send a text message to alert someone that Mr. Smith needs help.

Another test prototype is a special prompting robot designed to help people with dementia and their caregivers. One common issue with that population is that they are often high-functioning, but they might forget the steps involved in, say, making a cup of tea or brushing their teeth. The prompting robot (named Ed) is about the same size and shape as a water cooler, with a square monitor for a "head."

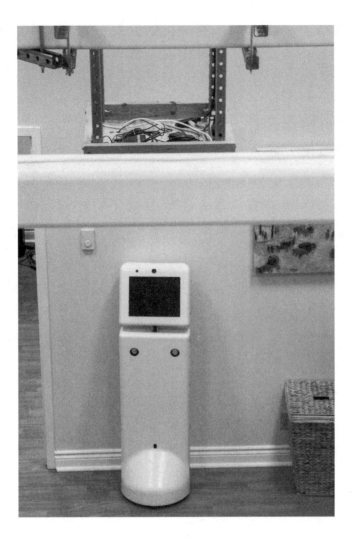

Prompting robot in the HomeLab living room.
Carol Svec

This moveable robot can provide step-by-step directions for accomplishing a number of tasks. Because it has a video screen and speakers, it can provide both audio and visual instructions. This type of robot helps not only the person with dementia but also his or her caregivers, who don't have to constantly prompt the instructions or do the tasks themselves.

Researchers are also working on a number of assistive devices, such as transfer poles to help people get out of bed in the morning, bathroom designs with slip-resistant surfaces and optimally positioned grab bars, handrail design, staircase improvements, and many other home improvements to prevent falls.

"We are working on smart homes of the future—ideally designed homes for seniors," says Dr. Campos. "But, realistically, people are living in homes that are sometimes one hundred years old, so we are also trying to figure out how to make what exists work, without major construction or costs or installation problems. We're looking for solutions that are easy and fluid and working with what you have."

It's one thing to understand how our ability to balance changes over time, what sensory signals need to get processed by the brain, and how the brain makes sense of them all—from vestibular, visual, auditory, proprioceptive, and kinesthetic systems. We can try to improve our own balance, build our muscle synergies, and exercise to keep ourselves as strong as possible. But those ideas don't help a ninety-year-old man with Parkinson's disease stay upright and healthy in his own home.

The Toronto Rehabilitation Institute is working both sides of that problem—understanding the way physiology and behavior affect balance and researching how to change the environment to make the world more user-friendly for older people. This type of research provides both the yin and the yang of balance, completing the circle and increasing the chances that we'll all be safer, now and in the future. As we age. And start to lose our balance.

"What I'm most excited about is what we're just beginning," says Dr. Campos. "Traditionally people have been working in silos with respect to looking at the effects of vision on balance, *or* the effects of proprioception or somatosensory input, *or* the effects of the vestibular system. Now we have the potential to quantify all those areas, to look at them together. And not just what's happening in healthy older adults, but understanding the role and the impact of sensory impairment in the context of dementia."

Again, closing the circle. That seems to be the theme of this institute.

Next Steps . . . on Ice

Imagine walking across an icy sidewalk after an inch of freezing rain has fallen on it. Each step feels precarious. You take tiny steps, resting periodically with a wide stance. But you have to get to the other side, so you continue, small step after small step, barely creeping along. Then your next step is a slide, and your feet come out from under you. You instinctively throw your hands up in an attempt to grasp at . . . nothing. But then, before you hit ground, your body comes to a halt—the harness suspended from the ceiling stops your fall.

That is one scenario routinely used in the CEAL, which consists of four laboratory spaces or pods—large, self-contained cubes, each set up for different research challenges. At the center of the huge basement lab is a giant twenty-foot-by-twenty-foot hydraulic motion platform that can be tilted in any direction. It has to be this big to support the individual laboratory pods, which can be lifted by crane and placed on the hydraulic platform. The interchangeable pods offer unique research environments.

The Toronto Rehab CEAL hydraulic motion platform in midtilt
with StairLab pod attached on top.
Sid Tabak Photography

For example, WinterLab features that ice-walking experience; the whole floor can be frozen—you can literally figure skate in that room. The technicians also can create real snow and wind gusts up to about nine miles per hour.* Research participants have had to walk across slippery, melting ice wearing different winter footwear, giving

---

*Basically everything that you need to conduct research relevant to Canada.

researchers information about how people balance on slick surfaces and about which footwear provides better grip on ice. Subjects' behaviors are documented by motion capture videos, eye trackers, and force plates, and—of course—each participant wears a harness for safety.*

Another pod, StreetLab, has a treadmill and large projection screens that allow researchers to simulate walking down the familiar main streets of Toronto. They can gather data about how sounds might distract a person and make his or her gait more unpredictable while crossing an intersection; or they can look at how a temporary disability, like a knee or hip replacement, affects the possibility of losing balance while walking. StairLab has a fully instrumented staircase—researchers can measure the body mechanics used to navigate the steps, including the force with which people step and how much pressure they apply to a handrail. The newest pod is DriveLab, a highly sophisticated, state-of-the-art driving simulator used to research driving performance of older people under challenging situations. It uses virtual reality technology to give the visual experience of driving, while controlling the external conditions by creating real glare and real rain in the simulator—actual water droplets falling on the windshield, so that you would have to use the wipers to get them off. And the steering simulator isn't a simulator at all; it is a real car—a fully equipped Audi A3 with a 360-degree virtual reality projection screen. The car has all normal functions on the inside. The only things missing are the tires and engine, which just added unnecessary weight. The car is mounted on a rotating platform, so it can simulate spinning out in a skid.†

---

*Toronto Rehab has created a website that lists the best footwear for winter, with a search option so you can see how your shoes or boots stack up. Check it out at www.ratemytreads.com.

†I have always wanted to take one of those defensive driving courses, where you learn how to come out of a spin or outmaneuver someone chasing you, but I'm too chicken about personal injury. I would love to do a virtual version of that.

Talking robots and a benevolent eye-in-the-sky; rooms that rain and snow; scientists having a direct influence on staircases country-wide—it's all very next-generation, a view into the future of balance research.

Oh, and let's not forget where we started: motion sickness. No matter how blue-sky balance research gets, it has its roots in motion sickness, which still affects millions of people. Dr. Keshavarz, research scientist at Toronto Rehab and adjunct professor in the Department of Psychology at Ryerson University, is doing some unique and surprising research on the subject. For example, he has looked at the effect of sound on motion sickness and—perhaps most unexpected—the effect of smells on motion sickness.* That's a real-world problem of a different sort, and one that I know well.† After so much time getting lost in the details of endolymph density, branches of the Achilles tendon, infrasound, and neuronal reflexes, talking about the mundane—but always fascinating—motion sickness feels like coming home.

As I said earlier—and I hope this has become clear throughout this book—we don't *have* a sense of balance; we *are* balance. Every day, we do what not even the most technologically advanced robot can do: we walk, stand, and chop carrots for a salad while balancing on two feet. Every muscle movement is a tribute to the brain's ability to synthesize and prioritize balance signals.

So what does it take to knock us off our feet? With everything I have learned, that feels like a Zen riddle, like *What is the sound of one hand clapping?* It takes almost nothing to lay us flat. A patch of ice, a banana peel, a pinecone under your heel can throw you

---

*In case you're wondering, initial work suggested that pleasant smells make motion sickness a little better. While the definition of *pleasant* is individual, the most common pleasant smell is roses.

†Have I told you about my time in the Vominator?

off center. A wayward thought can ruin a gymnast's balance beam routine, costing her a place on the Olympic team. And remember Helen, who has persistent postural-perceptual dizziness? She can be laid low by stirring a pot of soup. And yet, there are those of us who can maintain footing in a hurricane* or while fighting the symptoms of Parkinson's disease or coming back from total eradication of proprioceptive cues or when the vestibular system is damaged by a disease or medication. Even in zero gravity, some of us can maintain a sense of which way is up. We just need a little help as we age.

What does it take to knock us off our feet? Everything. Nothing. Or, as is the only appropriate answer to any Zen riddle: Ha!

---

*I see this during every hurricane, when TV stations send their weather reporters out into the whipping wind and rain.

Acknowledgments

THIS BOOK WOULDN'T EXIST without the inspiration, participation, and support of so many people that I'm afraid I'll miss thanking someone important. Any missing name is a mere oversight, not a snub.

This book was born after a particularly harsh New York City winter when sidewalks were best navigated with ice picks and crampons. My literary agent, Jane Dystel (of Dystel, Goderich & Bourret), called to say that she had been thinking about the mechanics of balance, and that was that. Jane gave me not only the inspiration but also the support I needed to bring this book to completion. Thank you, Jane!!!*

I also need to thank (!!!) several people at Chicago Review Press. My editor, Lisa Reardon, allowed me great latitude in content and offered compassion in timing after a particularly challenging year. She has a rare and enviable ability to guide without controlling. Project editor and detail wrangler Ellen Hornor has an almost frightening range of knowledge and the patience to put it to use.

---

*Jane hates exclamation points, but sometimes a multitude is the only proper way to convey depth of feeling.

Designer Debbie Berne found the precise right note in translating this book cover into visual art. Marketing manager Mary Kravenas helped me understand the big picture of how CRP moves words into readers' hands, and publicist Olivia Aguilar put that plan into action. And, finally, proofreader and cleaner-upper Sandra Smith made my day with LOLs and very few questions. Chicago Review Press is a unique and valuable publishing house that seems to employ only friendly and highly competent people. I appreciate them all.

This book would not have been possible without the individuals who told me their stories of dealing with balance problems, and the scientists and physicians who opened their labs and research files to me. Balance researchers are the most enthusiastic and passionate group of scientists I have ever met. All have been so generous with their time and knowledge and supremely tolerant of my questions. I feel privileged to have had a chance to experience their world.

I must put in a good word and hearty thanks to the National Association of Science Writers and to NASW executive director Tinsley Davis. This organization is a wonderful support in many ways, including providing access to research articles via ScienceDirect. I burned through hundreds of articles while writing this book.

A big hug and thank-you to my first reader, reference organizer, and sister, Ann Agrawal. Without her help and early comments, this book would be at least a year late.

I also owe so much to my mom and stepdad, Marina and Ted Rudisill. My mother had that horrendous fall I described in the introduction. I saw it happen, and I thought we had lost her. She survived, miraculously, without lasting brain injury. She, my stepdad, and sister live just around the corner. They fed us when my work lasted well past dinnertime, they watched the dogs when I needed to sequester myself, and they—as well as good friends Jill Kleinberg and Sharon Ayscue—are a never-ending source of love and encouragement.

Finally, thank you to my husband, Bill. He is the one who suffers for my art. He's the best. (It's in print, so it must be true!)

Selected References

General Books

Furman, Joseph M., and Stephen P. Cass, eds. *Balance Disorders: A Case-Study Approach*. Philadelphia: F. A. Davis, 1996.

Herdman, Susan J., and Richard A. Clendaniel, eds. *Vestibular Rehabilitation*. 4th edition. Philadelphia: F. A. Davis, 2014.

Chapter 1: Motion Sickness

Bonato, Frederick, Andrea Bubka, Shaziela Ishak, et al. "The Sickening Rug: A Repeating Static Pattern That Leads to Motion-Sickness-Like Symptoms." *Perception* 40, no. 4 (January 2011): 493–96.

Bubka, Andrea, Frederick Bonato, Scottie Urmey, et al. "Rotation Velocity Change and Motion Sickness in an Optokinetic Drum." *Aviation, Space, and Environmental Medicine* 77, no. 8 (September 2006): 811–15.

Bubka, Andrea, and Frederick Bonato. "Optokinetic Drum Tilt Hastens the Onset of Vection-Induced Motion Sickness." *Aviation, Space, and Environmental Medicine* 74, no. 4 (May 2003): 315–19.

Golding, John F. "Motion Sickness Susceptibility." *Autonomic Neuroscience* 129, no. 1–2 (November 2006): 67–76. doi:10.1016/j.autneu.2006.07.019.

Izumi-Kurotani, Akemi, Yoshihiro Mogami, Makoto Okuno, and Masamichi Yamashita. "Frog Experiment Onboard Space Station Mir." *Advances in Space Biology and Medicine* 6 (February 1997): 193–211.

Ji, Jennifer T., Richard H. So, and Raymond Cheung. "Isolating the Effects of Vection and Optokinetic Nystagmus on Optokinetic Rotation-Induced Motion Sickness." *Human Factors* 51, no. 5 (October 2009): 739–51.

Kennedy, Robert S., Julie Drexler, and Robert C. Kennedy. "Research in Visually Induced Motion Sickness." *Applied Ergonomics* 41, no. 4 (July 2010): 494–503. doi:10.1016/j.apergo.2009.11.006.

Keshavarz, Behrang, and Heiko Hecht. "Pleasant Music as a Countermeasure Against Visually Induced Motion Sickness." *Applied Ergonomics* 45, no. 3 (May 2014): 521–27. doi:10.1016/j.apergo.2013.07.009.

Keshavarz, Behrang, Lawrence J. Hettinger, Robert S. Kennedy, et al. "Demonstrating the Potential for Dynamic Auditory Stimulation to Contribute to Motion Sickness." Edited by M. S. Malmierca. *PLOS ONE* 9, no. 7 (2014): e101,016–e101,019. doi:10.1371/journal.pone.0101016.

Keshavarz, Behrang, Bernhard E. Riecke, Lawrence J. Hettinger, and Jennifer L. Campos. "Vection and Visually Induced Motion Sickness: How Are They Related?" *Frontiers in Psychology* 6 (April 20, 2015): 1–11. doi:10.3389/fpsyg.2015.00472.

Keshavarz, Behrang, Daniela Stelzmann, Aurora Paillard, and Heiko Hecht. "Visually Induced Motion Sickness Can Be Alleviated by Pleasant Odors." *Experimental Brain Research* 233, no. 5 (May 2015): 1,353–64.

Lackner, James R. "Motion Sickness: More than Nausea and Vomiting." *Experimental Brain Research* 232, no. 8 (August 2014): 2,493–510.

Matsangas, Panagiotis, and Michael E. McCauley. "Sopite Syndrome: A Revised Definition." *Aviation, Space, and Environmental Medicine* 85, no. 6 (June 2014): 672–73.

Reason, J. T. "Motion Sickness Adaptation: A Neural Mismatch Model." *Journal of the Royal Society of Medicine* 71, no. 11 (November 1978): 819–29.

Schmäl, F. "Neuronal Mechanisms and the Treatment of Motion Sickness." *Pharmacology* 91, no. 3–4 (April 2013): 229–41.

Tal, Dror, Peter Gilbey, Ronen Bar, and Avi Shupak. "Seasickness Pathogenesis and the Otolithic Organs: Vestibular Evoked Myogenic Potentials Study—Preliminary Results." *Israel Medicine Association Journal* 9, no. 9 (September 2007): 641–44.

Thornton, William E., and Frederick Bonato. "Space Motion Sickness and Motion Sickness: Symptoms and Etiology." *Aviation, Space, and Environmental Medicine* 84, no. 7 (July 2013): 716–21. doi:10.3357/ASEM.3449.2013.

Yamashita, Masamichi, Akemi Izumi-Kurotani, Yoshihiro Mogami, et al. "The Frog in Space (FRIS) Experiment Onboard Space Station Mir: Final Report and Follow-on Studies." *Biological Sciences in Space* 11, no. 4 (December 1997): 313–20.

Chapters 2 and 3: Anatomy

Angelaki, Dora E., and Kathleen E. Cullen. "Vestibular System: The Many Facets of a Multimodal Sense." *Annual Review of Neuroscience* 31, no. 1 (February 2008): 125–50.

Aretaeus. "De causis et signis acutorum morborum." *The Extant Works of Aretaeus, the Cappadocian.* Edited by Francis Adams LL.D. Boston: Milford House, 1972. Originally published in 1856. Perseus Digital Library. Accessed October 27, 2015.

Bradshaw, Andrew P., Ian S. Curthoys, Michael J. Todd, et al. "A Mathematical Model of Human Semicircular Canal Geometry: A New Basis for Interpreting Vestibular Physiology." *Journal of the Association for Research in Otolaryngology* 11, no. 2 (June 2010): 145–59. doi:10.1007/s10162-009-0195-6.

Daocai, Wang, Wang Qing, Wang Ximing, et al. "Size of the Semicircular Canals Measured by Multidetector Computed Tomography in Different Age Groups." *Journal of Computer Assisted Tomography* 38, no. 2 (March 2014): 196–209. doi:10.1097/RCT.0b013e3182aaf21c.

Eatock, Ruth Anne, and Jocelyn E. Songer. "Vestibular Hair Cells and Afferents: Two Channels for Head Motion Signals." *Annual Review of Neuroscience* 34, no. 1 (July 2011): 501–34. doi:10.1146/annurev-neuro-061010-113710.

Gunn, John C. *Dr. Gunn's New Family Physician Home Book of Health.* 230th edition. New York: Saalfield, 1901.

Hain, Timothy C. "Neurophysiology of Vestibular Rehabilitation." *NeuroRehabilitation* 29, no. 2 (October 2011): 127–41. doi:10.3233/NRE-2011-0687.

Hashimoto, Shinichiro, Hideaki Naganuma, Koji Tokumasu, et al. "Three-Dimensional Reconstruction of the Human Semicircular Canals and Measurement of Each Membranous Canal Plane Defined by Reid's Stereotactic Coordinates." *Annals of Otology, Rhinology & Laryngology* 114, no. 12 (December 2005): 934–38.

Heidrenreich, Katherine D., Kelly Beaudoin, and Judith A. White. "Cervicogenic Dizziness as a Cause of Vertigo While Swimming: An Unusual Case Report." *American Journal of Otolaryngology* 29, no. 6 (November–December 2008): 429–31.

Helminski, Janet Odry, Imke Janssen, Despina Kotaspouikis, et al. "Strategies to Prevent Recurrence of Benign Paroxysmal Positional Vertigo." *Archives of Otolaryngology Head and Neck Surgery* 131, no. 4 (April 2005): 344–48. doi:10.1001/archotol.131.4.344.

Hunt, William T., Eleanor F. Zimmerman, and Malcolm P. Hilton. "Modifications of the Epley (Canalith Repositioning) Manoeuvre for Posterior Canal Benign Paroxysmal Positional Vertigo (BPPV)." *Cochrane Database of Systematic Reviews* 4, no. 4 (April 18, 2012). doi: 10.1002/14651858. CD008675.pub2.

Ju-Young, Lee, Kang-Jae Shin, Jeong-Nam Kim, et al. "A Morphometric Study of the Semicircular Canals Using Micro-CT Images in Three-Dimensional Reconstruction." *Anatomical Record* 296, no. 5 (May 2013): 834–39.

Juul-Kristensen, Birgit, Brian Clausen, Inge Ris, et al. "Increased Neck Muscle Activity and Impaired Balance Among Females with Whiplash-Related Chronic Neck Pain: A Cross-Sectional Study." *Journal of Rehabilitation Medicine* 45, no. 4 (April 2013): 376–84.

Kaski, Diego, and Adolfo M. Bronstein. "Epley and Beyond: An Update on Treating Positional Vertigo." *Practical Neurology* 14, no. 4 (August 2014): 210–21. doi:10.1136/practneurol-2013-000690.

Khan, Sarah, Richard Chang. "Anatomy of the Vestibular System: A Review." *NeuroRehabilitation* 32, no.3 (May 2013): 437–43. doi:10.3233/NRE-130866.

Lackner, James R., and Paul DiZio. "Vestibular, Proprioceptive, and Haptic Contributions to Spatial Orientation." *Annual Review of Psychology* 56, no. 1 (January 2005): 115–47. doi:10.1146/annurev.psych.55.090902.142023.

Lopez-Escamez, Jose A., John Carey, Won-Ho Chung, et al. "Diagnostic Criteria for Menière's Disease." *Journal of Vestibular Research* 25, no. 1 (January 2015): 1–7. doi:10.3233/VES-150549.

Pearce, J. M. "Benign Paroxysmal Vertigo, and Bárány's Caloric Reactions." *European Neurology* 57, no. 4 (April 2007): 246–48. doi:10.1159/000101292.

Purves, Dale, George J. Augustine, David Fitzpatrick, et al. *Neuroscience.* 5th edition. Sunderland, MA: Sinauer Associates, 2012.

Stewart, Thomas Grainger. *Lectures on Giddiness and on Hysteria in the Male.* 2nd edition. Edinburgh and London: Young J. Pentland, 1898.

Wade, Nicholas J. "The Original Spin Doctors: The Meeting of Perception and Insanity." *Perception* 34, no. 3 (March 2005): 253–60.

Chapter 4: Proprioception

Balslev, Daniela, Jonathan Cole, and R. Christopher Miall. "Proprioception Contributes to the Sense of Agency During Visual Observation of Hand Movements: Evidence from Temporal Judgments of Action." *Journal of Cognitive Neuroscience* 19, no. 9 (September 2007): 1,535–41.

Cole, Jonathan, and Ian Waterman. *Pride and a Daily Marathon*. Cambridge, MA: MIT Press, 1995.

McNeill, David, Liesbet Quaeghebeur, and Susan Duncan. "IW—'The Man Who Lost His Body.'" In *Handbook of Phenomenology and Cognitive Sciences*, edited by Shaun Gallagher and Daniel Schmicking. Dordrecht, Netherlands: Springer, 2010.

Yousif, Nada, Jonathan Cole, John Rothwell, and Jörn Diedrichsen. "Proprioception in Motor Learning: Lessons from a Deafferented Subject." *Experimental Brain Research* 233, no. 8 (August 2015): 2,449–59.

Chapter 5: Self-Orientation—Up

Angelaki, Dora E., and Kathleen E. Cullen. "Vestibular System: The Many Facets of a Multimodal Sense." *Annual Review of Neuroscience* 31, no. 1 (January 2008): 125–50. doi:10.1146/annurev.neuro.31.060407.125555.

Bisdorff, Alexandre R., C. J. Wolsley, Dimitri Anastasopoulos, et al. "The Perception of Body Verticality (Subjective Postural Vertical) in Peripheral and Central Vestibular Disorders." *Brain* 119, no. 5 (October 1996): 1,523–34.

Clark, Torin K., Michael C. Newman, Charles M. Oman, et al. "Modeling Human Perception of Orientation in Altered Gravity." *Frontiers in Systems Neuroscience* 9 (May 5, 2015): 1–13. doi:10.3389/fnsys.2015.00068.

Harris, Laurence R., Michael J. Carnevale, Sarah D'Amour, et al. "How Our Body Influences Our Perception of the World." *Frontiers in Psychology* 6 (June 12, 2015): 257–310. doi:10.3389/fpsyg.2015.00819.

Harris, Laurence R., Rainer Herpers, Thomas Hofhammer, et al. "How Much Gravity Is Needed to Establish the Perceptual Upright?" Edited by M. Longo. *PLOS ONE* 9, no. 9 (November 2014): e106,207. doi:10.1371/journal.pone.0106207.

Harris, Laurence R., Michael R. Jenkin, Heather L. Jenkin, et al. "Where's the Floor?" *Seeing and Perceiving* 23, no.1 (June 2010): 81–88. doi:10.1163/187847510X490826.

Hoover, Adria E., and Laurence R. Harris. "The Role of the Viewpoint on Body Ownership." *Experimental Brain Research* 233, no. 4 (December 2014): 1,053–60. doi:10.1007/s00221-014-4181-9.

Jenkin, Heather L., Richard T. Dyde, James E. Zacher, et al. "The Relative Role of Visual and Non-Visual Cues in Determining the Perceived Direction of 'Up': Experiments in Parabolic Flight." *Acta Astronautica* 56, no. 9–12 (May 2005): 1,025–32. doi:10.1016/j.actaastro.2005.01.030.

Jenkin, Michael R., Richard T. Dyde, Heather L. Jenkin, et al. "Perceptual Upright: The Relative Effectiveness of Dynamic and Static Images Under Different Gravity States." *Seeing and Perceiving* 24, no. 1 (February 2011): 53–64. doi:10.1163/187847511X555292.

Lackner, James R., and Paul DiZio. "Vestibular, Proprioceptive, and Haptic Contributions to Spatial Orientation." *Annual Review of Psychology* 56, no. 1 (January 2005): 115–47. doi:10.1146/annurev.psych.55.090902.142023.

Lackner, James R., and A. Graybiel. "Parabolic Flight: Loss of Sense of Orientation." *Science* 206, no. 4,422 (November 1979): 1,105–8.

van Erp, Jan B., and Hendrik A. van Veen. "Touch Down: The Effect of Artificial Touch Cues on Orientation in Microgravity." *Neuroscience Letters* 404, no. 1–2 (August 2006): 78–82.

Chapter 6: PPPD

Bronstein, Adolfo M. "The Visual Vertigo Syndrome." *Acta Oto-Laryngologica Supplement* 520, part 1 (January 1995): 45–48.

Bronstein, Adolfo, M. "Visual Symptoms and Vertigo." *Neurologic Clinics* 23, no. 3 (August 2005): 705–13.

Bronstein, Adolfo M., John F. Golding, and Michael A. Gresty. "Vertigo and Dizziness from Environmental Motion: Visual Vertigo, Motion Sickness, and Drivers' Disorientation." *Seminars in Neurology* 33, no. 3 (July 2013): 219–30.

Chang, Chih-Pei, and Timothy C. Hain. "A Theory for Treating Dizziness Due to Optical Flow (Visual Vertigo)." *CyberPsychology & Behavior* 11, no. 4 (August 2008): 495–98.

Guerraz, Michel, Lucy Yardley, P. Bertholon, et al. "Visual Vertigo: Symptom Assessment, Spatial Orientation and Postural Control." *Brain* 124, part 8 (August 2001): 1,646–56.

Keshavarz, Behrang, Bernhard Riecke, Lawrence J. Hettinger, and Jennifer L. Campos. "Vection and Visually Induced Motion Sickness: How Are They Related?" *Frontiers in Psychology* 6, no. 472 (April 20, 2015): 1–11.

Sawle, Guy. "Visual Vertigo." *Lancet* 347, no. 9,007 (April 13, 1996): 986–87.

Whitney, Susan L., Patrick J. Sparto, James R. Cook, et al. "Symptoms Elicited in Persons with Vestibular Dysfunction while Performing Gaze Movements in Optic Flow Environments." *Journal of Vestibular Research* 23, no. 1 (April 2013): 51–60.

Chapter 7: Sound

Al'tman, Ya A., Victor S. Gurfinkel, O.V. Varyagina, and Yu S. Levik. "The Effects of Moving Sound Images on Postural Responses and the Head Rotation Illusion in Humans." *Neuroscience and Behavioral Physiology* 35, no. 1 (January 2005): 103–6.

Feldmann, Joachim, and F. A. Pitten. "Effects of Low Frequency Noise on Man—A Case Study." *Noise and Health* 7, no. 25 (October–December 2004): 23–28.

Leventhall, Geoff. "What Is Infrasound?" *Progress in Biophysics and Molecular Biology* 93, no. 1–3 (January–April 2007): 130–37.

Mikołajczak, J., Sylwester Borowski, J. Marć-Pieńkowska, et al. "Preliminary Studies on the Reaction of Growing Geese (Anseranser f. domestica) to the Proximity of Wind Turbines." *Polish Journal of Veterinary Sciences* 16, no. 4 (November 2013): 1–8.

Rumalla, Kavelin, Adham M. Karim, and Timothy E. Huller. "The Effect of Hearing Aids on Postural Stability." *Laryngoscope* 125, no. 3 (March 2015): 720–23.

Salt, Alec N., and Timothy E. Hullar. "Responses of the Ear to Low Frequency Sounds, Infrasound and Wind Turbines." *Hearing Research* 268, no. 1–2 (September 1, 2010): 12–21.

Schomer, Paul D., John Erdreich, Pranav K. Pamidighantam, James H. Boyle. "A Theory to Explain Some Physiological Effects of the Infrasonic Emissions at Some Wind Farm Sites." *Journal of the Acoustical Society of America* 137, no. 3 (March 2015): 1,356–65.

Chapter 8: Alcohol & Pharmacology

Boyle, Nichola, Vasi Naganathan, and Robert G. Cumming. "Medication and Falls: Risk and Optimization." *Clinics in Geriatric Medicine* 26 (November 2010): 583–605.

Chen, Ying, Ling-Ling Zhu, and Quan Zhou. "Effects of Drug Pharmacokinetic/ Pharmacodynamics Properties, Characteristics of Medication Use, and Relevant Pharmacological Interventions on Fall Risk in Elderly Patients." *Therapeutics and Clinical Risk Management* 10 (June 13, 2014): 437–48.

Chimirri, Serafina, Rossana Aiello, Carmela Mazzitello, et al. "Vertigo/Dizziness as a Drug's Adverse Reaction." *Journal of Pharmacology and Pharmacotherapeutics* 4, no. 5 (December 2013): 104–6. doi:10.4103/0976-500X.120969.

Darlington, Cynthia L., and Paul F. Smith. "Vestibulotoxicity Following Aminoglycoside Antibiotics and Its Prevention." *Current Opinion in Investigational Drugs* 4, no. 7 (July 2003): 841–46.

Guthrie, O'neil W. "Aminoglycoside Induced Ototoxicity." *Toxicology* 249, no. 2–3 (July 2008): 91–96. doi:10.1016/j.tox.2008.04.015.

Hafstrom, Anna, Mitesh Patel, Fredrik Modig, et al. "Acute Alcohol Intoxication Impairs Segmental Body Alignment in Upright Standing." *Journal of Vestibular Research* (July 2014): 1–8. doi:10.3233/VES-140513.

Lin, Emerald, and Kathy Aligene. "Pharmacology of Balance and Dizziness." *NeuroRehabilitation* 32, no. 3 (April 2013): 529–42. doi:10.3233/NRE-130875.

Sedó-Cabezón, Lara, Pere Boadas-Vaello, Carla Soler-Martín, and Jordi Llorens. "Vestibular Damage in Chronic Ototoxicity: A Mini-Review." *NeuroToxicology* 43 (July 2014): 21–27. doi:10.1016/j.neuro.2013.11.009.

Shoair, Osama A., Abner N. Nyandege, and Patricia W. Slattum. "Medication-Related Dizziness in the Older Adult." *Otolaryngolic Clinics of North America* 44, no. 2 (March 2011): 455–71. doi:10.1016/j.otc.2011.01.014.

Chapters 9 and 10: Cybersickness and Virtual Reality

Benoit, Michel, Rachid Guerchouche, Pierre-David Petit, et al. "Is It Possible to Use Highly Realistic Virtual Reality in the Elderly? A Feasibility Study with Image-Based Rendering." *Neuropsychiatric Disease and Treatment*, no. 11 (March 3, 2015): 557–63.

Bergeron, Mathieu, Catherine L. Lortie, and Matthieu J. Guitton. "Use of Virtual Reality Tools for Vestibular Disorders Rehabilitation: A Comprehensive Analysis." *Advances in Medicine* 2015, no. 3 (November 2015): 1–9. doi:10.1155/2015/916735.

Bisson, Etienne J., B. Contant, Heidi Sveistrup, and Yves Lajoie. "Functional Balance and Dual-Task Reaction Times in Older Adults Are Improved by Virtual Reality and Biofeedback Training." *CyberPsychology & Behavior* 10, no. 1 (February 2007): 16–23. doi:10.1089/cpb.2006.9997.

Bonato, Frederick, Andrea Bubka, and Stephen Palmisano. "Combined Pitch and Roll and Cybersickness in a Virtual Environment." *Aviation, Space, and Environmental Medicine* 80, no. 11 (October 2009): 941–45.

Cho, Ki Hun, Kyoung Jin Lee, and Chang Ho Song. "Virtual-Reality Balance Training with a Video-Game System Improves Dynamic Balance in Chronic Stroke Patients." *Tohoku Journal of Experimental Medicine* 228, no. 1 (August 2012): 69–74. doi:10.1620/tjem.228.69.

Crosbie, Jacqueline H., Sheila Lennon, Michael D. J. McNeill, and Suzanne M. McDonough. "Virtual Reality in the Rehabilitation of the Upper Limb after Stroke: The User's Perspective." *CyberPsychology & Behavior* 9, no. 2 (April 2006): 137–41. doi:10.1089/cpb.2006.9.137.

David, Daniel, Silviu Matu, and Oana Alexandra David. "New Directions in Virtual Reality-Based Therapy for Anxiety Disorders." *International Journal of Cognitive Therapy* 6, no. 2 (May 2013): 114–37. doi:10.1521/ijct.2013.6.2.114.

Duque, Gustavo, Derek Boersma, Griselda Loza-Diaz, et al. "Effects of Balance Training Using a Virtual-Reality System in Older Fallers." *Clinical Interventions in Aging* 8 (February 2013): 257. doi:10.2147/CIA.S41453.

Fung, Joyce, Carol L. Richards, Francine Malouin, et al. "A Treadmill and Motion Coupled Virtual Reality System for Gait Training Post-Stroke." *CyberPsychology & Behavior* 9, no. 2 (May 2006): 157–62.

Furman, Joseph M., Susan L. Whitney, and Patrick J. Sparto. "Perceived Anxiety and Simulator Sickness in a Virtual Grocery Store in Persons with and Without Vestibular Dysfunction." (March 2015): 1–7.

Jerald, Jason. *The VR Book: Human-Centered Design for Virtual Reality*. San Rafael, CA: Morgan & Claypool, 2015.

Jerald, Jason, Mary Whitton, and Frederick Phillips Brooks Jr. "Scene-Motion Thresholds During Head Yaw for Immersive Virtual Environments." *ACM Transactions on Applied Perception* 9, no. 1 (February 2012): 1–23. doi:10.1145/2134203.2134207.

Johnson, David M. *Research Report 1832: Introduction to and Review of Simulator Sickness Research*. U.S. Army Research Institute for the Behavioral and Social Sciences: April 2005. http://www.dtic.mil/dtic/tr/fulltext/u2/a434495.pdf.

Kennedy, Robert S., Jennifer E. Fowlkes, Kevin S. Berbaum, and Michael G. Lilienthal. "Use of a Motion Sickness History Questionnaire for Prediction of Simulator Sickness." *Aviation, Space, and Environmental Medicine* 63, no. 7 (July 1992): 588–93.

Kennedy, Robert S., Norman E. Lane, Kevin S. Berbaum, and Michael G. Lilienthal. "Simulator Sickness Questionnaire: An Enhanced Method for Quantifying Simulator Sickness." *International Journal of Aviation Psychology* 3, no. 3 (July 1993): 203–20. doi:10.1207/s15327108ijap0303_3.

Kennedy, Robert S., Michael G. Lilienthal, Kevin S. Berbaum, et al. "Simulator Sickness in U.S. Navy Flight Simulators." *Aviation, Space, and Environmental Medicine* 60, no. 1 (January 1989): 10–16.

Keshner, Emily A., Jefferson Streepey, Yasin Dhaher, and Timothy Hain. "Pairing Virtual Reality with Dynamic Posturography Serves to Differentiate Between Patients Experiencing Visual Vertigo." *Journal of NeuroEngineering and Rehabilitation* 4 (July 2007): 24. doi:10.1186/1743-0003-4-24.

Kim, Young Youn, Hyun Ju Kim, Eun Nam Kim, et al. "Characteristic Changes in the Physiological Components of Cybersickness." *Psychophysiology* 42, no. 5 (September 2005): 616–25. doi:10.1111/j.1469-8986.2005.00349.x.

Kuttuva, Manjuladevi, Rares Florin Boian, Alma Merians, et al. "The Rutgers Arm, a Rehabilitation System in Virtual Reality: A Pilot Study." *CyberPsychology & Behavior* 9, no. 2 (April 2006): 148–52. doi:10.1089/cpb.2006.9.148.

Lo, W. T., and Richard H. So. "Cybersickness in the Presence of Scene Rotational Movements along Different Axes." *Applied Ergonomics* 32, no. 1 (February 2001): 1–14.

Mejía-Rentería, Hernán D., and Iván J. Núñez-Gil. "Takotsubo Syndrome: Advances in the Understanding and Management of an Enigmatic Stress Cardiomyopathy." *World Journal of Cardiology* 8, no. 7 (July 2016): 413. doi:10.4330/wjc.v8.i7.413.

Meldrum, Dara, Susan Herdman, Roisin C. Vance, et al. "Effectiveness of Conventional Versus Virtual Reality Based Vestibular Rehabilitation in the Treatment of Dizziness, Gait and Balance Impairment in Adults with Unilateral Peripheral Vestibular Loss: A Randomised Controlled Trial." *Archives of Physical Medicine and Rehabilitation* 96, no. 7 (July 2015): 1,319–28.e1. doi: 10.1016/j.apmr.2015.02.032.

Miller, J. W., and J. E. Goodson. "Motion Sickness in a Helicopter Simulator." *Aerospace Medicine* (April 1960): 204–12.

Murray, Craig D., Emma Patchick, Stephen Pettifer, et al. "Immersive Virtual Reality as a Rehabilitative Technology for Phantom Limb Experience: A Protocol." *CyberPsychology & Behavior* 9, no. 2 (April 2006): 167–70. doi:10.1089/cpb.2006.9.167.

Naqvi, Syed Ali, Nasreen Badruddin, Aamir S. Malik, et al. "Does 3D Produce More Symptoms of Visually Induced Motion Sickness?" *Conference Proceedings: The Annual International Conference of the IEEE Engineering in Medicine and Biology Society* (July 2013): 6,405–8. doi:10.1109/EMBC.2013.6611020.

Nyberg, Lars, Lillemor Lundin-Olsson, Björn Sondell, et al. "Using a Virtual Reality System to Study Balance and Walking in a Virtual Outdoor Environment: A Pilot Study." *CyberPsychology & Behavior* 9, no. 4 (August 2006): 388–95.

Palmisano, Stephen, Robert S. Allison, Juno Kim, and Frederick Bonato. "Simulated Viewpoint Jitter Shakes Sensory Conflict Accounts of Vection." *Seeing and Perceiving* 24, no. 2 (March 2011): 173–200. doi:10.1163/187847511X570817.

Park, Junhyuck, and Jongeun Yim. "A New Approach to Improve Cognition, Muscle Strength, and Postural Balance in Community-Dwelling Elderly with a

3-D Virtual Reality Kayak Program." *Tohoku Journal of Experimental Medicine* 238, no. 1 (January 2016): 1–8. doi:10.1620/tjem.238.1.

Slobounov, Semyon, Elena Slobounov, and Karl Newell. "Application of Virtual Reality Graphics in Assessment of Concussion." *CyberPsychology & Behavior* 9, no. 2 (April 2006): 188–91. doi:10.1089/cpb.2006.9.188.

Solimini, Angelo G. "Are There Side Effects to Watching 3D Movies? A Prospective Crossover Observational Study on Visually Induced Motion Sickness." *PLOS ONE* 8, no. 2 (February 2013): e56,160–68. doi:10.1371/journal.pone.0056160.

Solimini, Angelo G., Alice Mannocci, Domitilla Di Thiene, and Giuseppe La Torre. "A Survey of Visually Induced Symptoms and Associated Factors in Spectators of Three Dimensional Stereoscopic Movies." *BMC Public Health* 12, no. 1 (September 2012): 779. doi:10.1186/1471-2458-12-779.

Tanaka, Nobuhisa, and Hideyuki Takagi. "Virtual Reality Environment Design of Managing Both Presence and Virtual Reality Sickness." *Journal of Physiological Anthropology and Applied Human Science* 23, no. 6 (December 2004): 313–17.

Taylor, Montoya, Anish Amin, and Charles Bush. "Three-Dimensional Entertainment as a Novel Cause of Takotsubo Cardiomyopathy." *Clinical Cardiology* 34, no. 11 (November 2011): 678–80. doi:10.1002/clc.20950.

Virk, S., and K. M. McConville. "Virtual Reality Applications in Improving Postural Control and Minimizing Falls." *Conference Proceedings: The Annual International Conference of the IEEE Engineering in Medicine and Biology Society* 1 (January 2006): 2,694–97.

Young, Sean D., Bernard D. Adelstein, and Stephen R. Ellis. "Demand Characteristics in Assessing Motion Sickness in a Virtual Environment: Or Does Taking a Motion Sickness Questionnaire Make You Sick?" *IEEE Transactions on Visualization and Computer Graphics* 13, no. 3 (April 2007): 422–28.

Chapter 11: Psychology

Altenmüller, Eckart, Christos I. Ioannou, and Andre Lee. "Apollo's Curse: Neurological Causes of Motor Impairments in Musicians." *Progress in Brain Research* 217 (February 19, 2015): 89–106.

Beilock, Sian L., and Thomas H. Carr. "On the Fragility of Skilled Performance: What Governs Choking Under Pressure?" *Journal of Experimental Psychology: General* 130, no. 4 (December 2001): 701–25.

Causer, Joe, Paul S. Holmes, Nick Smith, and A. Mark Williams. "Anxiety, Movement Kinematics, and Visual Attention in Elite-Level Performers." *Emotion* (June 2011): 595–602.

Cooke, Andrew M., Maria Kavussanu, David Mcintyre, and Chris Ring. "Psychological, Muscular and Kinematic Factors Mediate Performance under Pressure." *Psychophysiology* 47, no. 6 (October 2010): 1,109–18. doi: 10.1111/j.14 69-8986.2010.01021.x.

DeCaro, Marci S., Robin D. Thomas, Neil B. Albert, and Sian L. Beilock. "Choking under Pressure: Multiple Routes to Skill Failure." *Journal of Experimental Psychology: General* 140, no. 3 (May 2011): 390–406. doi: 10.1037/a0023466.

Dhungana, Samish, and Joseph Jankovic. "Yips and Other Movement Disorders in Golfers." *Movement Disorders* 28, no. 5 (April 2013): 576–81. doi: 10.1002/mds.25442.

Durlik, Caroline, Flavia Cardini, and Manos Tsakiris. "Being Watched: The Effect of Social Self-Focus on Interoceptive and Exteroceptive Somatosensory Perception." *Consciousness and Cognition* 25 (March 2014): 42–50. doi: 10.1016/j.concog.2014.01.010.

Gucciardi, Daniel F., Jay-Lee Longbottom, Ben Jackson, and James A. Dimmock. "Experienced Golfers' Perspectives on Choking under Pressure." *Journal of Sport and Exercise Psychology* 32, no. 1 (January 2010): 61–83.

Jankovic, Joseph, and Aidin Ashoori. "Movement Disorders in Musicians." *Movement Disorders* 23, no. 14 (October 2008): 1,957–65. doi: 10.1002/mds.22255.

Otten, Mark. "Choking vs. Clutch Performance: A Study of Sport Performance under Pressure." *Journal of Sport Exercise Psychology* 31, no. 5 (September 2009): 583–601.

Schücker, Linda, Norbert Hagemann, and Bernd Strauss. "Attentional Processes and Choking under Pressure." *Perceptual and Motor Skills* 116, no. 2 (March 2013): 671–89.

Smith, Aynsley M., Charles H. Adler, Debbie Crews, et al. "The 'Yips' in Golf." *Sports Medicine* 33, no. 1 (January 2003): 13–31.

Smith, Aynsley M., Susan A. Malo, Edward R. Laskowski, et al. "A Multidisciplinary Study of the 'Yips' Phenomenon in Golf: An Exploratory Analysis." *Sports Medicine* 30, no. 6 (December 2000): 423–37.

Yoshie, Michiko, Kazutoshi Kudo, Takayuki Murakoshi, and Tatsuyuki Ohtsuki. "Music Performance Anxiety in Skilled Pianists: Effects of Social-Evaluative Performance Situation on Subjective, Autonomic, and Electromyographic Reactions." *Experimental Brain Research* 199, no. 2 (September 2009): 117–26. doi: 10.1007/s00221-009-1979-y.

Yoshie, Michiko, Yoko Nagai, Hugo D. Critchley, and Neil A. Harrison. "Why I Tense Up When You Watch Me: Inferior Parietal Cortex Mediates an Audience's Influence on Motor Performance." *Scientific Reports* 6 (January 20, 2016): 19,305. doi: 10.1038/srep19305.

Chapter 12: Biomechanics

Collins, Steven H., M. Bruce Wiggin, and Gregory S. Sawicki. "Reducing the Energy Cost of Human Walking Using an Unpowered Exoskeleton." *Nature* 522, no. 7,555 (March 2015): 212–15. doi: 10.1038/nature14288.

Franz, Jason R. "The Age-Associated Reduction in Propulsive Power Generation in Walking." *Exercize and Sport Sciences Reviews* 44, no. 4 (June 2016): 129–36.

Takahashi, Kota Z., Michael D. Lewek, and Gregory S. Sawicki. "A Neuromechanics-Based Powered Ankle Exoskeleton to Assist Walking Post-Stroke: A Feasibility Study." *Journal of NeuroEngineering and Rehabilitation* 12, no. 1 (February 2015): 23.

Wang, Yang, and Manoj Srinivasan. "Stepping in the Direction of the Fall: The Next Foot Placement Can Be Predicted from Current Upper Body State in Steady-State Walking." *Biology Letters* 10, no. 9 (September 2014).

Chapters 13 and 14: Fall Risk

Bourane, Steeve, KS Grossmann, O Britz, et al. "Identification of a Spinal Circuit for Light Touch and Fine Motor Control." *Cell* 160, no. 3 (January 2015): 503–15.

Chen, Tiffany L., Tapomayukh Bhattacharjee, J. Lucas McKay, et al. "Evaluation by Expert Dancers of a Robot That Performs Partnered Stepping Via Haptic Interaction." *PLOS One* 10, no. 5 (May 2015): e0125,179.

Chvatal, Stacie A., and Lena H. Ting. "Common Muscle Synergies for Balance and Walking." *Frontiers in Computational Neuroscience* 7, no. 48 (May 2013): 1–14.

Francis, Carrie Anne, Jason R. Franz, Shawn M. O'Connor, and Darryl G. Thelen. "Gait Variability in Healthy Old Adults Is More Affected by a Visual Perturbation than by a Cognitive or Narrow Step Placement Demand." *Gait & Posture* 42, no. 3 (July 2015): 380–85.

Franz, Jason R. "The Age-Associated Reduction in Propulsive Power Generation in Walking." *Exercise and Sport Sciences Reviews* 44, no. 4 (July 2016): 129–36.

Franz, Jason R., Carrie Anne Francis, Mathew S. Allen, et al. "Advanced Age Brings a Greater Reliance on Visual Feedback to Maintain Balance During Walking." *Human Movement Science* 40 (April 2015): 381–92.

Franz, Jason R., Carrie Anne Francis, Matt Allen, and Darryl G. Thelen. "Visuomotor Entrainment and the Frequency-Dependent Response of Walking Balance to Perturbations." *IEEE Transactions on Neural Systems and Rehabilitation Engineering* (August 26, 2016): EPUB ahead of print.

Iwasaki, Shinichi, and Tatsuya Yamasoba. "Dizziness and Imbalance in the Elderly: Age-Related Decline in the Vestibular System." *Aging and Disease* 6, no. 1 (January 2015): 38–47.

Jacobson, Gary P., Devin L. McCaslin, Sarah L. Grantham, and Erin G. Piker. "Significant Vestibular System Impairment Is Common in a Cohort of Elderly Patients Referred for Assessment of Falls Risk." *Journal of the American Academy of Audiology* 19, no. 10 (October 2008): 799–807.

Jilk, D. Joseph, Seyed A. Safavynia, and Lena H. Ting. "Contribution of Vision to Postural Behaviors During Continuous Support-Surface Translations." *Experimental Brain Research* 232, no. 1 (January 2014): 169–80.

Mansfield, Avril, Amy L. Peters, Barbara A. Lieu, and Brian Edward Maki. "Effect of a Perturbation-Based Balance Training Program on Compensatory Stepping and Grasping Reactions in Older Adults: A Randomized Controlled Trial." *Physical Therapy* 90, no. 4 (April 2010): 476–91.

Mansfield, Avril, Jennifer S. Wong, Jessica Bryce, et al. "Does Perturbation-Based Balance Training Prevent Falls? Systematic Review and Meta-Analysis of Preliminary Randomized Controlled Trials." *Physical Therapy* 95, no. 5 (May 2015): 700–709.

Moncada, Lainie Van Vost. "Management of Falls in Older Persons: A Prescription for Prevention." *American Family Physician* 84, no. 11 (November 2011): 1,267–76.

Park, Junhyuck, and Jongeun Yim. "A New Approach to Improve Cognition, Muscle Strength, and Postural Balance in Community-Dwelling Elderly with a 3-D Virtual Reality Kayak Program." *Tohoku Journal of Experimental Medicine* 238, no. 1 (January 2016): 1–8.

Ramkhalawansingh, Robert, Behrang Keshavarz, Bruce Haycock, et al. "Age Differences in Visual-Auditory Self-Motion Perception During a Simulated Driving Task." *Frontiers in Psychology* 7, no. 595 (April 2016): 1–12.

Index

Italicized page numbers refer to images